国家林业和草原局普通高等教育"十四五"规划教材

林业国际文书导读

娄瑞娟 李 芝 主编

中国林业出版社
China Forestry Publishing House

内 容 简 介

《林业国际文书导读》着眼于生态环境保护和可持续发展，选择了与人类可持续发展和森林多功能利用相关的 8 个联合国文书，聚焦了全球环境治理和森林经营利用等热点问题，旨在拓宽高校学生国际视野，培养科研素养，增强创新意识，提高创新能力，促进学生建立关注世界、服务社会的责任感和使命感，助力学生成为引领林业发展和创新的高素质、复合型、国际化农林人才，增强他们在各自学科领域内的国际竞争力和学术话语权。本教材语言原真，内容贴近现实，针对林业院校学科特点和学生未来职业需求而编写，具有很强的可读性和实用性；术语、背景注解简明扼要，思考练习题设计紧凑，是高等农林院校培养国际性专业人才，以及政府和相关外事管理人员的必备参考用书。

图书在版编目（CIP）数据

林业国际文书导读/娄瑞娟，李芝主编. —北京：中国林业出版社，2022.12
国家林业和草原局普通高等教育"十四五"规划教材
ISBN 978-7-5219-2048-2

Ⅰ. ①林… Ⅱ. ①娄… ②李… Ⅲ. ①林业—国际交流—高等学校—教材 Ⅳ. ①S7-12

中国版本图书馆 CIP 数据核字（2022）第 254357 号

责任编辑：高红岩　王奕丹
责任校对：苏　梅
封面设计：睿思视界视觉设计

出版发行：中国林业出版社
　　　　　（100009，北京市西城区刘海胡同 7 号，电话 83223120）
电子邮箱：cfphzbs@163.com
网　　址：www.forestry.gov.cn/lycb.html
印　　刷：北京中科印刷有限公司
版　　次：2022 年 12 月第 1 版
印　　次：2022 年 12 月第 1 次印刷
开　　本：787mm×1092mm　1/16
印　　张：13.625
字　　数：335 千字
定　　价：46.00 元

《林业国际文书导读》编写人员

主　　编　娄瑞娟　李　芝
副 主 编　杜景芬　由　华　朱红梅
编写人员　（按姓氏笔画排序）
　　　　　　王雪梅（北京林业大学）
　　　　　　王景枝（中国传媒大学）
　　　　　　由　华（北京林业大学）
　　　　　　申　通（中国林业科学研究院）
　　　　　　白雪莲（北京林业大学）
　　　　　　朱红梅（北京林业大学）
　　　　　　李　芝（北京林业大学）
　　　　　　李　辉（北京外国语大学）
　　　　　　杜景芬（北京林业大学）
　　　　　　杨惠媛（北京林业大学）
　　　　　　罗凌志（北京林业大学）
　　　　　　娄瑞娟（北京林业大学）
　　　　　　凌舒亚（北京林业大学）

前 言
Preface

　　新文科背景下创新型农林人才培养模式强调研究性教学，注重个性化培养，拓宽国际化视野，着力提升学生的创新意识、创新能力和科研素养。双一流高校要加快推进拔尖创新型农林人才培养方案建设，培养一批引领农林业创新发展的高层次、高水平、复合型高素质农林人才，这是农林类高校在国家培养卓越农林人才教育培养计划中肩负的重要时代责任。卓越农林人才的培养离不开农林学科领域专业国际化知识。培养一流的国际化人才必须具备的关键素质包括宽广的国际化视野、扎实的本专业国际化知识、熟悉和掌握国际惯例、较强的跨文化沟通能力和独立的国际活动能力等。

　　在经济全球化和高等教育国际化的新时代，英语已成为世界上各学科成果交流和科技经济发展报告的国际通用语。国际化人才应该具备的素质离不开高水平的外语能力，尤其是专业领域内的英语交流能力。培养新时代的农林高校大学生，使之在各自的学科领域内具有国际竞争力和国际话语权是农林类高校新的历史使命。为了提升学生的国际视野，助力实现这一人才培养目标，我们组织相关团队编写了《林业国际文书导读》。本着"前瞻意识、思政引领、创新能力、学用结合"的理念，本教材以国际涉林文书为依托，着眼于生态环境保护和可持续发展的国际社会热点，引导学生养成关注世界、服务社会的责任感和使命感。本教材共 8 章，内容涉及全球可持续发展和林业保护等多方面内容，分别涵盖《联合国人类环境会议宣言》《21 世纪议程》《联合国千年宣言》《联合国森林文书》《巴黎协定》《改变我们的世界：2030 年可持续发展议程》《2015 年后国际森林安排决议》和《联合国森林战略规划（2017—2030 年）》。

　　针对林业院校学科特点和学生未来职业需求，本教材提供了英语读译语言技能的学习和训练，帮助学生了解和掌握相关语言知识和基本技能，提高涉外交流能力。本教材的编写依托农林类高校的绿色事业基础，培养生态文明建设思想内涵，融合工具性和人文性，注重培养家国情怀和文化素养，促进学生知行合一；助力学生用英语讲好中国绿色故事，提升学生的英语思辨能力，构建思政引领的融合型教学模式。

　　本教材适用范围广，实用性强，尤其对林学和环境保护类专业学生以及相关从业人员了解和掌握林业国际文书方面的相关知识大有助益，能够帮助学习者提高参与国际交流和处理国际事务的能力。本教材也适合政府管理人员、技术人员、国际交流人员和翻译人员参考使用。

　　本教材系教育部新文科研究与改革实践项目"产出导向与持续质量改进模式下的新文科外语类课程体系和教材体系建设与实践研究"（项目编号：2021070015）和北京林业大学教育教学改革与研究重点项目"多层次，模块化，链条式大学英语人才培养课程体系建设"（BJFU2022JYZD012）的阶段性成果。感谢北京林业大学教务处和中国林业出版社的鼎力支持！本教材编写过程还得到了中国林业科学研究院资源信息研究所王宏所长的指导，并得到了兄弟院校同行的大力支持。在此一并表示感谢。

  由于水平有限，不当之处在所难免，恳请同行、专家和读者不吝赐教，以期使之日臻完善。

编 者

2022 年 10 月

目 录

Contents

前 言

第 1 章　联合国人类环境会议宣言
Chapter 1　Declaration of the United Nations Conference on the Human Environment ··· 1

第 2 章　21 世纪议程
Chapter 2　Agenda 21 ·· 21

第 3 章　联合国千年宣言
Chapter 3　United Nations Millennium Declaration ·· 52

第 4 章　联合国森林文书
Chapter 4　United Nations Forest Instrument ·· 75

第 5 章　巴黎协定
Chapter 5　Paris Agreement ·· 98

第 6 章　改变我们的世界：2030 年可持续发展议程
Chapter 6　Transforming Our World: the 2030 Agenda for Sustainable Development ···· 129

第 7 章　2015 年后国际森林安排决议
Chapter 7　International Arrangement on Forests beyond 2015 ····························· 157

第 8 章　联合国森林战略规划（2017—2030 年）
Chapter 8　United Nations Strategic Plan for Forests 2017—2030 ························· 183

第 1 章　联合国人类环境会议宣言

Chapter 1　Declaration of the United Nations Conference on the Human Environment

Background and Significance

《联合国人类环境会议宣言》是 1972 年第一次全球环境会议的成果，又称《斯德哥尔摩人类环境会议宣言》，简称《斯德哥尔摩宣言》或《人类环境宣言》（以下简称《宣言》）。《宣言》提出了 7 点共同看法和 26 条共同遵守的原则，涵盖了人的环境权利和保护环境的义务，保护和合理利用各种自然资源，防治污染，促进经济和社会发展，使发展同保护和改善环境协调一致，援助发展中国家，对发展和保护环境进行计划和规划，实行适当的人口政策，发展环境科学、技术和教育，销毁核武器和其他一切大规模杀伤性武器，加强国家对环境的管理，促进国际合作等。《宣言》本身不具有法律约束力，但是由于它反映了国际社会的共同信念，对国际环境法的发展产生了深远的影响，主要表现在：

- 《宣言》第一次概括了《国际环境法》的原则和规则，其中一些还成了后来国际环境条约有约束力的原则和规则。
- 尽管这些原则和规则没有法律约束力，但为国际环境保护提供了政治和道义上所应遵守的规范。
- 为各国制定和发展本国国内环境法提供了可遵循和借鉴的原则和规则。

历史证明，《宣言》是一份具有划时代意义的历史文献，标志着人类对环境问题认识的转折，开创了人类社会环境保护事业的新纪元，是人类环境保护史上的第一座里程碑。

Text Study

Declaration of the United Nations Conference on the Human Environment

The United Nations Conference on the Human Environment,

Having met at Stockholm from 5 to 16 June 1972,

Having considered the need for a common outlook and for common principles to inspire and guide the peoples of the world in the preservation and enhancement of the human environment.

I

Proclaims that:

1. Man is both creature and moulder of his environment, which gives him physical sustenance and affords him the opportunity for intellectual, moral, social and spiritual growth. In the long and tortuous evolution of the human race on this planet a stage has been reached when, through the rapid acceleration of science and technology, man has acquired the power to transform his environment in countless ways and on an unprecedented scale. Both aspects of man's environment, the natural and the man-made, are essential to his well-being and to the enjoyment of basic human rights—even the right to life itself.

2. The protection and improvement of the human environment is a major issue which affects the well-being of peoples and economic development throughout the world; it is the urgent desire of the peoples of the whole world and the duty of all Governments.

3. Man has constantly to sum up experience and go on discovering, inventing, creating and advancing. In our time, man's capability to transform his surroundings, if used wisely, can bring to all peoples the benefits of development and the opportunity to enhance the quality of life. Wrongly or heedlessly applied, the same power can do incalculable harm to human beings and the human environment. We see around us growing evidence of man-made harm in many regions of the earth: dangerous levels of pollution in water, air, earth and living beings; major and undesirable disturbances to the ecological balance of the biosphere; destruction and depletion of irreplaceable resources; and gross deficiencies, harmful to the physical, mental and social health of man, in the man-made environment, particularly in the living and working environment.

4. In the developing countries most of the environmental problems are caused by under-development. Millions continue to live far below the minimum levels required for a decent human existence, deprived of adequate food and clothing, shelter and education, health and sanitation. Therefore, the developing countries must direct their efforts to development, bearing in mind their priorities and the need to safeguard and improve the environment. For the same purpose, the industrialized countries should make efforts to reduce the gap themselves and the developing countries. In the industrialized countries, environmental problems are generally related to industrialization and technological development.

5. The natural growth of population continuously presents problems for the preservation of the environment, and adequate policies and measures should be adopted, as appropriate, to face these problems. Of all things in the world, people are the most precious. It is the people that propel social progress, create social wealth, develop science and technology and, through their hard work, continuously transform the human environment. Along with social progress and the advance of production, science and technology, the capability of man to improve the environment increases with each passing day.

6. A point has been reached in history when we must shape our actions throughout the

world with a more prudent care for their environmental consequences. Through ignorance or indifference we can do massive and irreversible harm to the earthly environment on which our life and well-being depend. Conversely, through fuller knowledge and wiser action, we can achieve for ourselves and our posterity a better life in an environment more in keeping with human needs and hopes. There are broad vistas for the enhancement of environmental quality and the creation of a good life. What is needed is an enthusiastic but calm state of mind and intense but orderly work. For the purpose of attaining freedom in the world of nature, man must use knowledge to build, in collaboration with nature, a better environment. To defend and improve the human environment for present and future generations has become an imperative goal for mankind—a goal to be pursued together with, and in harmony with, the established and fundamental goals of peace and of worldwide economic and social development.

7. To achieve this environmental goal will demand the acceptance of responsibility by citizens and communities and by enterprises and institutions at every level, all sharing equitably in common efforts. Individuals in all walks of life as well as organizations in many fields, by their values and the sum of their actions, will shape the world environment of the future. Local and national governments will bear the greatest burden for large-scale environmental policy and action within their jurisdictions. International cooperation is also needed in order to raise resources to support the developing countries in carrying out their responsibilities in this field. A growing class of environmental problems, because they are regional or global in extent or because they affect the common international realm, will require extensive cooperation among nations and action by international organizations in the common interest. The Conference calls upon Governments and peoples to exert common efforts for the preservation and improvement of the human environment, for the benefit of all the people and for their posterity.

II
Principles

States the common conviction that:

Principle 1

Man has the fundamental right to freedom, equality and adequate conditions of life, in an environment of a quality that permits a life of dignity and well-being, and he bears a solemn responsibility to protect and improve the environment for present and future generations. In this respect, policies promoting or perpetuating apartheid, racial segregation, discrimination, colonial and other forms of oppression and foreign domination stand condemned and must be eliminated.

Principle 2

The natural resources of the earth, including the air, water, land, flora and fauna and especially representative samples of natural ecosystems, must be safeguarded for the benefit of present and future generations through careful planning or management, as appropriate.

Principle 3

The capacity of the earth to produce vital renewable resources must be maintained and, wherever practicable, restored or improved.

Principle 4

Man has a special responsibility to safeguard and wisely manage the heritage of wildlife and its habitat, which are now gravely imperilled by a combination of adverse factors. Nature conservation, including wildlife, must therefore receive importance in planning for economic development.

Principle 5

The non-renewable resources of the earth must be employed in such a way as to guard against the danger of their future exhaustion and to ensure that benefits from such employment are shared by all mankind.

Principle 6

The discharge of toxic substances or of other substances and the release of heat, in such quantities or concentrations as to exceed the capacity of the environment to render them harmless, must be halted in order to ensure that serious or irreversible damage is not inflicted upon ecosystems. The just struggle of the peoples of all countries against pollution should be supported.

Principle 7

States shall take all possible steps to prevent pollution of the seas by substances that are liable to create hazards to human health, to harm living resources and marine life, to damage amenities or to interfere with other legitimate uses of the sea.

Principle 8

Economic and social development is essential for ensuring a favorable living and working environment for man and for creating conditions on earth that are necessary for the improvement of the quality of life.

Principle 9

Environmental deficiencies generated by the conditions of under-development and natural disasters pose grave problems and can best be remedied by accelerated development through the transfer of substantial quantities of financial and technological assistance as a supplement to the domestic effort of the developing countries and such timely assistance as may be required.

Principle 10

For the developing countries, stability of prices and adequate earnings for primary commodities and raw materials are essential to environmental management, since economic factors as well as ecological processes must be taken into account.

Principle 11

The environmental policies of all States should enhance and not adversely affect the present or future development potential of developing countries, nor should they hamper the attainment of better living conditions for all, and appropriate steps should be taken by States and international organizations with a view to reaching agreement on meeting the possible national and international economic consequences resulting from the application of environmental measures.

Principle 12

Resources should be made available to preserve and improve the environment, taking into account the circumstances and particular requirements of developing countries and any costs which may emanate from their incorporating environmental safeguards into their development planning and the need for making available to them, upon their request, additional international technical and financial assistance for this purpose.

Principle 13

In order to achieve a more rational management of resources and thus to improve the environment, States should adopt an integrated and coordinated approach to their development planning so as to ensure that development is compatible with the need to protect and improve environment for the benefit of their population.

Principle 14

Rational planning constitutes an essential tool for reconciling any conflict between the needs of development and the need to protect and improve the environment.

Principle 15

Planning must be applied to human settlements and urbanization with a view to avoiding adverse effects on the environment and obtaining maximum social, economic and environmental benefits for all. In this respect projects which are designed for colonialist and racist domination must be abandoned.

Principle 16

Demographic policies which are without prejudice to basic human rights and which are deemed appropriate by Governments concerned should be applied in those regions where the rate of population growth or excessive population concentrations are likely to have adverse effects on the environment of the human environment and impede development.

Principle 17

Appropriate national institutions must be entrusted with the task of planning, managing or controlling the environmental resources of States with a view to enhancing environmental quality.

Principle 18

Science and technology, as part of their contribution to economic and social development, must be applied to the identification, avoidance and control of environmental risks and the solution of environmental problems and for the common good of mankind.

Principle 19

Education in environmental matters, for the younger generation as well as adults, giving due consideration to the underprivileged, is essential in order to broaden the basis for an enlightened opinion and responsible conduct by individuals, enterprises and communities in protecting and improving the environment in its full human dimension. It is also essential that mass media of communications avoid contributing to the deterioration of the environment, but, on the contrary, disseminates information of an educational nature on the need to project and improve the environment in order to enable man to develop in every respect.

Principle 20

Scientific research and development in the context of environmental problems, both national and multinational, must be promoted in all countries, especially the developing countries. In this connection, the free flow of up-to-date scientific information and transfer of experience must be supported and assisted, to facilitate the solution of environmental problems; environmental technologies should be made available to developing countries on terms which would encourage their wide dissemination without constituting an economic burden on the developing countries.

Principle 21

States have, in accordance with the *Charter of the United Nations* and the principles of international law, the sovereign right to exploit their own resources pursuant to their own environmental policies, and the responsibility to ensure that activities within their jurisdiction or control do not cause damage to the environment of other States or of areas beyond the limits of national jurisdiction.

Principle 22

States shall cooperate to develop further the international law regarding liability and compensation for the victims of pollution and other environmental damage caused by activities within the jurisdiction or control of such States to areas beyond their jurisdiction.

Principle 23

Without prejudice to such criteria as may be agreed upon by the international community, or to standards which will have to be determined nationally, it will be essential in all cases to consider the systems of values prevailing in each country, and the extent of the applicability of standards which are valid for the most advanced countries but which may be inappropriate and of unwarranted social cost for the developing countries.

Principle 24

International matters concerning the protection and improvement of the environment should be handled in a cooperative spirit by all countries, big and small, on an equal footing. Cooperation through multilateral or bilateral arrangements or other appropriate means is essential to effectively control, prevent, reduce and eliminate adverse environmental effects resulting from activities conducted in all spheres, in such a way that due account is taken of the sovereignty and interests of all States.

Principle 25

States shall ensure that international organizations play a coordinated, efficient and dynamic role for the protection and improvement of the environment.

Principle 26

Man and his environment must be spared the effects of nuclear weapons and all other means of mass destruction. States must strive to reach prompt agreement, in the relevant international organs, on the elimination and complete destruction of such weapons.

21st plenary meeting
16 June 1972

第 1 章 联合国人类环境会议宣言
Chapter 1 Declaration of the United Nations Conference on the Human Environment

Notes

1. The United Nations Conference on the Human Environment

为应对日益严峻的环境的挑战，来自 113 个国家和地区的政府代表和民间人士于 1972 年 6 月 5 日至 16 日在瑞典首都斯德哥尔摩召开了联合国人类环境会议，就当代世界环境问题以及保护全球环境战略等问题进行了商讨，会议经过 12 天的讨论交流通过了著名的《联合国人类环境会议宣言》(又称《斯德哥尔摩宣言》或《人类环境宣言》)、《人类环境行动计划》和其他若干建议和决议。会议呼吁各国政府和人民为保护和改善人类环境，造福全世界人民，造福子孙后代而共同努力，提出了响遍世界的环境保护口号：只有一个地球！本次会议是人类历史上第一次以环境问题为主题而召开的国际会议。在这次会议上，国际社会第一次规定了人类对全球环境的权利与义务的共同原则，标志着人类共同环境保护历程的开始。环境问题自此列入国际议事日程。会议还确立每年的 6 月 5 日为"世界环境日"（World Environment Day）。

1972 年，在举行第一次人类环境会议前，中国代表团积极参与了该宣言的起草工作，并在会上共享了经周恩来总理提出并审定的中国政府关于环境保护的 32 字方针，即"全面规划，合理布局，综合利用，化害为利，依靠群众，大家动手，保护环境，造福人民"，鲜明地表达了中国政府对环境问题的观点和重视，受到了大会的尊重。时任燃料化学工业部副部长的唐克同志率领中国代表团出席了本次会议，并于 6 月 10 日上午在联合国人类环境会议的全体会议上发言。他指出，维护和改善人类环境，是关系到世界各国人民生活和经济发展的一个重要问题，中国政府和人民积极支持与赞助这个会议。他还阐述了中国代表团在维护和改善人类环境问题上的主张，揭露了有人无视超级大国大量制造和储存核武器，却不加区分地反对一切核试验的伪善行径，并且重申了中国政府关于全面禁止和彻底销毁核武器的一贯主张。

2. Charter of the United Nations

《联合国宪章》是联合国的基本大法，于 1945 年 6 月 26 日在旧金山会议上签署，1945 年 10 月 24 日正式生效。1945 年 4 月 25 日，美国旧金山，包括中国在内的 50 个国家的 288 名代表，抱着"欲免后世再遭今代人类两度身历惨不堪言之战祸"的决心，参加了这次联合国国际组织会议。经过为期两个月的商讨，与会代表一致通过了《联合国宪章》。1945 年 6 月 26 日，举行了《联合国宪章》签字仪式，中国代表第一个在宪章的中文、法文、俄文、英文、西文（拉丁文）5 种联合国正式语言文件上签字，随后是法国、苏联、英国、美国 4 国代表依次签字，最后签字的是与会的其他 45 个国家；同年 10 月 24 日《联合国宪章》开始生效，联合国正式成立。包括后来补签的波兰在内的 51 个国家成为联合国创始会员国。在签字的中国代表中，有一位中国共产党人，他就是参加过中共一大、走过长征路的中国共产党创始人之一的董必武。

《联合国宪章》除序言和结语外，共分 19 章 111 条，不仅确立了联合国的宗旨、原则和组织机构设置，还规定了成员国的责任、权利和义务，以及处理国际关系、维护世界和平与安全的基本原则和方法。《联合国宪章》规定，联合国的宗旨是"维护国际和平及安全""制止侵略行为""发展国际间以尊重各国人民平等权利自决原则为基础的友好关系"和"促成国际合作"等；还规定了联合国及其成员国应遵循各国主权平等，以和平方式解决国际争端,在国际关系中不使用武力或武力威胁来侵犯他国领土完整和主权,以及联合国不得干涉各国内政等原则。这些基本准则符合世界各国人民的根本利益,在今天仍然具有重要的现实意义,为维护世界的稳定与安宁发挥着不可或缺的作用。

Key Words and Phrases

1. amenity /əˈmiːnəti/ n. (usually pl.) things that are provided for people's convenience, enjoyment, or comfort 生活便利设施

2. apartheid /əˈpɑːtaɪt/ n. a political system in South Africa in which people were divided into racial groups and kept apart by law (南非曾经施行的) 种族隔离制

3. conversely /ˈkɒnvɜːsli/ adv. (formal) in a way that is the opposite or reverse of sth. 相反地;反过来

4. conviction /kənˈvɪkʃn/ n. a strong opinion or belief 坚定的看法(或信念)

5. deficiency /dɪˈfɪʃ(ə)nsi/ n. the state of not having, or not having enough of, sth. that is essential 缺乏;缺少;不足

6. demographic /ˌdeməˈɡræfɪk/ adj. relating to or concerning the statistics relating to the people who live there 人口(学)的;人口统计(学)的

7. depletion /dɪˈpliːʃn/ n. the act of decreasing something markedly 损耗;耗减

8. deterioration /dɪˌtɪərɪəˈreɪʃ(ə)n/ n. process of changing to an inferior state 恶化;衰退

9. emanate /ˈeməneɪt/ vt. (formal) to produce or show sth. 产生;表现;显示
   vi. (~ from) proceed or issue forth, as from a source 来自(于);(从…)散发出

10. dissemination /dɪˌsemɪˈneɪʃ(ə)n/ n. the act of distributing information or knowledge so that it reaches many people or organizations 传播;宣传

11. halt /hɔːlt/ v. to stop; to make sb. or sth. stop (使)停止,停下

12. hamper /ˈhæmpə(r)/ vt. to prevent sb. from easily doing or achie-

			ving sth. 妨碍；阻止
13. heedlessly	/ˈhiːdləsli/	adv.	without care or concern 掉以轻心地
14. impede	/ɪmˈpiːd/	v.	[often passive] (formal) to delay or stop the progress of sth. 阻碍；阻止
15. imperil	/ɪmˈperəl/	v.	(formal) to put sth. or sb. in danger 使陷于危险；危及
16. irreversible	/ˌɪrɪˈvɜːsəb(ə)l/	adj.	incapable of being changed back to what it was before 不可逆转的；无法复原（或挽回）的
17. jurisdiction	/ˌdʒʊərɪsˈdɪkʃn/	n.	an area or a country in which a particular system of laws has authority 管辖区域；管辖范围 (law) the right and power to interpret and apply the law 司法权；审判权；管辖权
18. moulder	/ˈməʊldə(r)/	n.	(US molder) a person or thing that shapes or influences other people or things in a particular way 塑造者
19. perpetuate	/pəˈpetʃueɪt/	vt.	(formal) to make sth. such as a bad situation, a belief, etc. continue for a long time 使延续；使永久化
20. posterity	/pɒˈsterəti/	n.	(formal) all the people who will live in the future 后代；后裔；子孙
21. propel	/prəˈpel/	vt.	to move, drive or push sth. forward or in a particular direction 推动；驱动；推进
22. prudent	/ˈpruːd(ə)nt/	adj.	careful and sensible 谨慎的；慎重的；精明的
23. realm	/relm/	n.	an area of activity, interest, or knowledge 领域，范围 （formal）a country that has a king or queen 王国
24. reconcile	/ˈrekənsaɪl/	vt.	to find an acceptable way of dealing with two or more ideas, needs, etc. that seem to be opposed to each other 使和谐一致；调和；使配合
25. sanitation	/ˌsænɪˈteɪʃ(ə)n/	n.	the process of keeping places clean and healthy, especially by providing a sewage system and a clean water supply 公共卫

生；环境卫生

26. segregation	/ˌsegrɪˈgeɪʃn/	n.	the act or policy of separating people of different races, religions or sexes and treating them in a different way （对不同种族、宗教或性别的人所采取的）隔离并区别对待，隔离政策
27. sovereign	/ˈsɒvrɪn/	adj.	[only before noun] (of a country or state) free to govern itself; completely independent 有主权的；完全独立的
		n.	(formal) a king or a queen 君主；元首
28. sustenance	/ˈsʌstənəns/	n.	(formal) the food and drink that people, animals and plants need to live and stay healthy 食物；营养；养料
29. tortuous	/ˈtɔːtʃuəs/	adj.	very long and complicated 转弯抹角的；绕圈子的；迂回复杂的
30. unprecedented	/ʌnˈpresɪdentɪd/	adj.	that has never happened, been done or been known before 前所未有的；空前的；没有先例的
31. unwarranted	/ʌnˈwɒrəntɪd/	adj.	(formal) not reasonable or necessary; not appropriate 不合理的；不必要的；无正当理由的；不适当的
32. vista	/ˈvɪstə/	n.	a view from a particular place, especially a beautiful view from a high place.（尤指从高处看到的）景色
			(formal) a range of things that might happen in the future（未来可能发生的）一系列情景，一连串事情

33. be compatible with   与…和谐相处；与…相配的
34. in all walks of life   各行各业
35. in collaboration with   与…合作
36. on an equal footing   公平地；平等地
37. on an unprecedented scale   以空前的规模

Exercises

Exercise 1 Reading Comprehension

Directions: *Read the first part of **Declaration of the United Nations Conference on the Human Environment**, and decide whether the following statements are true or false. Write T*

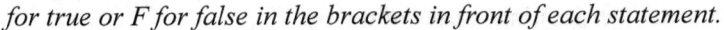

for true or F for false in the brackets in front of each statement.

1. (   ) Man's environment, including not only the natural one but also the artificial one, plays a vital role in his well-being and in the enjoyment of basic human rights.

2. (   ) Human beings are capable of shaping their environment in a manner either desirable or destructive, so they should check their power and think carefully before making any change to their surroundings.

3. (   ) Since most of the environmental problems facing developing countries are caused by under-development, they can make every attempt to develop first without considering the environment.

4. (   ) The developed countries should take into account narrowing down the gap between themselves and the developing countries as well as developing industry and technology.

5. (   ) Owing to massive and irreversible harm done to the earth, it is impossible for man to achieve a better life in an environment harmonious with his needs and hopes.

6. (   ) To achieve the environmental goal, local and national governments apart from making environmental policies should commit themselves to raising resources to support the developing countries.

Exercise 2   Skimming and Scanning

Directions: *Read the following passage excerpted from the **Declaration of the United Nations Conference on the Human Environment**. At the end of the passage, there are six statements. Each statement contains information given in one of the paragraphs of the passage. Identify the paragraph from which the information is derived. Each paragraph is marked with a letter. You may choose a paragraph more than once. Answer the questions by writing the corresponding letter in the brackets in front of each statement.*

The United Nations Conference on the Human Environment states the common conviction that:

A) Man has the fundamental right to freedom, equality and adequate conditions of life, in an environment of a quality that permits a life of dignity and well-being, and he bears a solemn responsibility to protect and improve the environment for present and future generations. In this respect, policies promoting or perpetuating apartheid, racial segregation, discrimination, colonial and other forms of oppression and foreign domination stand condemned and must be eliminated.

B) The natural resources of the earth, including the air, water, land, flora and fauna and especially representative samples of natural ecosystems, must be safeguarded for the benefit of present and future generations through careful planning or management, as appropriate.

C) The non-renewable resources of the earth must be employed in such a way as to guard against the danger of their future exhaustion and to ensure that benefits from such employment are shared by all mankind.

D) States shall take all possible steps to prevent pollution of the seas by substances that are liable to create hazards to human health, to harm living resources and marine life, to damage amenities or to interfere with other legitimate uses of the sea.

E) The environmental policies of all States should enhance and not adversely affect the present or future development potential of developing countries, nor should they hamper the attainment of better living conditions for all, and appropriate steps should be taken by States and international organizations with a view to reaching agreement on meeting the possible national and international economic consequences resulting from the application of environmental measures.

F) Resources should be made available to preserve and improve the environment, taking into account the circumstances and particular requirements of developing countries and any costs which may emanate- from their incorporating environmental safeguards into their development planning and the need for making available to them, upon their request, additional international technical and financial assistance for this purpose.

G) Science and technology, as part of their contribution to economic and social development, must be applied to the identification, avoidance and control of environmental risks and the solution of environmental problems and for the common good of mankind.

H) States shall ensure that international organizations play a coordinated, efficient and dynamic role for the protection and improvement of the environment.

I) Man and his environment must be spared the effects of nuclear weapons and all other means of mass destruction. States must strive to reach prompt agreement, in the relevant international organs, on the elimination and complete destruction of such weapons.

1. (   ) To prevent the seas from being polluted, measures should be taken to eliminate substances detrimental to human health, living resources and marine life.

2. (   ) When making environmental policies, all countries should take into consideration how to help developing countries develop well.

3. (   ) An environment of quality is essential for man to enjoy his basic rights and live with dignity, so he should take the responsibility to protect and improve the environment.

4. (   ) Rational use of the non-renewable resources means not only their being free from exhaustion in the future but also their bringing benefits to all human beings.

5. (   ) To save man and his environment from being ruined by weapons of mass destruction, a universal agreement should be reached on their disposal and destruction.

6. (   ) Policies resulting in racial inequality, prejudice, exploitation and other forms of injustice should be denounced and removed.

Exercise 3   Word Formation

Directions: *In this section, there are ten sentences from the* **Declaration of the United Nations Conference on the Human Environment**. *You are required to complete these sentences with the proper form of the words given in the blanks.*

1. Man is both creature and moulder of his environment, which gives him physical sustenance and affords him the opportunity for _____, moral, social and spiritual growth. (intellect)

2. Wrongly or heedlessly applied, the same power can do _____ harm to human beings and the human environment. (calculate)

第1章 联合国人类环境会议宣言
Chapter 1 Declaration of the United Nations Conference on the Human Environment

3. We see around us growing evidence of man-made harm in many regions of the earth: dangerous levels of pollution in water, air, earth and living beings; major and undesirable disturbances to the ecological balance of the biosphere; destruction and depletion of _____ resources. (replace)

4. There are broad vistas for the _____ of environmental quality and the creation of a good life. (enhance)

5. A growing class of environmental problems, because they are regional or global in extent or because they affect the common international realm, will require _____ cooperation among nations and action by international organizations in the common interest. (extend)

6. Planning must be applied to human settlements and _____ with a view to avoiding adverse effects on the environment and obtaining maximum social, economic and environmental benefits for all. (urban)

7. In this respect projects which are designed for _____ and racist domination must be abandoned. (colony)

8. Education in environmental matters, for the younger generation as well as adults, giving due consideration to the_____, is essential in order to broaden the basis for an enlightened opinion and responsible conduct by individuals, enterprises and communities in protecting and improving the environment in its full human dimension. (privilege)

9. States have, in accordance with the *Charter of the United Nations* and the principles of international law, the sovereign right to exploit their own resources_____ to their own environmental policies, and the responsibility to ensure that activities within their jurisdiction or control do not cause damage to the environment of other States or of areas beyond the limits of national jurisdiction. (pursue)

10. Cooperation through multilateral or bilateral arrangements or other appropriate means is essential to effectively control, prevent, reduce and eliminate adverse environmental effects resulting from activities conducted in all spheres, in such a way that due account is taken of the _____ and interests of all States. (sovereign)

Exercise 4　Translation
Section A
Directions: Read ***Declaration of the United Nations Conference on the Human Environment*** and complete the sentences by translating into English the Chinese given in the blanks.

1. The protection and improvement of the human environment is a major issue which affects _____(人民的福祉及经济的发展) throughout the world.

2. Millions continue to live far below the minimum levels required for a decent human existence, deprived of adequate food and clothing, _____ (居所、教育、保健及卫生设施).

3. Along with social progress and the advance of production, science and technology, _____ (人类改善环境的能力与日俱增).

4. _____ (地球生产重要的可再生资源的能力) must be

maintained and, wherever practicable, restored or improved.

5. Man has a special responsibility to safeguard and wisely manage _____ _____ (野生生物遗产及其栖息地), which are now gravely imperilled by a combination of adverse factors.

6. The discharge of toxic substances or of other substances and the release of heat, in such quantities or concentrations as to exceed the capacity of the environment to render them harmless, must be halted in order to ensure that _____ (生态系统免遭严重的或不可挽回的损失).

7. Environmental deficiencies generated by the conditions of under-development and natural disasters pose grave problems and can best be remedied by accelerated development through _____ (调动大量的财政及技术援助) as a supplement to the domestic effort of the developing countries and such timely assistance as may be required.

8. For the developing countries, stability of prices and adequate earnings for primary commodities and raw materials are essential to environmental management, since _____ (经济因素和生态过程) must be taken into account.

9. Planning must be applied to human settlements and urbanization with a view to avoiding adverse effects on the environment and obtaining_____ (最大化的社会、经济、环境效益) for all.

10. States have, in accordance with the *Charter of the United Nations* and the principles of international law, the sovereign right to _____ (按自己国家的环境政策开发利用自己的自然资源), and the responsibility to ensure that activities within their jurisdiction or control do not cause damage to the environment of other States or of areas beyond the limits of national jurisdiction.

Section B

Directions: *Translate the following sentences from English into Chinese.*

1. Man is both creature and moulder of his environment, which gives him physical sustenance and affords him the opportunity for intellectual, moral, social and spiritual growth. In the long and tortuous evolution of the human race on this planet a stage has been reached when, through the rapid acceleration of science and technology, man has acquired the power to transform his environment in countless ways and on an unprecedented scale. Both aspects of man's environment, the natural and the man-made, are essential to his well-being and to the enjoyment of basic human rights— the right to life itself. (*Proclamation 1*)

2. Man has constantly to sum up experience and go on discovering, inventing, creating and advancing. In our time, man's capability to transform his surroundings, if used wisely, can bring to all peoples the benefits of development and the opportunity to enhance the quality of life. Wrongly or heedlessly applied, the same power can do incalculable harm to human beings and the human environment. We see around us growing evidence of man-made harm in many regions of the earth: dangerous levels of pollution in water, air, earth and living beings; major and undesirable disturbances to the ecological balance of the biosphere; destruction and

depletion of irreplaceable resources; and gross deficiencies, harmful to the physical, mental and social health of man, in the man-made environment, particularly in the living and working environment. (*Proclamation 3*)

3. A point has been reached in history when we must shape our actions throughout the world with a more prudent care for their environmental consequences. Through ignorance or indifference we can do massive and irreversible harm to the earthly environment on which our life and well being depend. Conversely, through fuller knowledge and wiser action, we can achieve for ourselves and our posterity a better life in an environment more in keeping with human needs and hopes. There are broad vistas for the enhancement of environmental quality and the creation of a good life. What is needed is an enthusiastic but calm state of mind and intense but orderly work. For the purpose of attaining freedom in the world of nature, man must use knowledge to build, in collaboration with nature, a better environment. To defend and improve the human environment for present and future generations has become an imperative goal for mankind—a goal to be pursued together with, and in harmony with, the established and fundamental goals of peace and of worldwide economic and social development. (*Proclamation 6*)

4. To achieve this environmental goal will demand the acceptance of responsibility by citizens and communities and by enterprises and institutions at every level, all sharing equitably in common efforts. Individuals in all walks of life as well as organizations in many fields, by their values and the sum of their actions, will shape the world environment of the future. Local and national governments will bear the greatest burden for large-scale environmental policy and action within their jurisdictions. International cooperation is also needed in order to raise resources to support the developing countries in carrying out their responsibilities in this field. A growing class of environmental problems, because they are regional or global in extent or because they affect the common international realm, will require extensive cooperation among nations and action by international organizations in the common interest. The Conference calls upon Governments and peoples to exert common efforts for the preservation and improvement of the human environment, for the benefit of all the people and for their posterity. (*Proclamation 7*)

5. Demographic policies which are without prejudice to basic human rights and which are deemed appropriate by Governments concerned should be applied in those regions where the rate of population growth or excessive population concentrations are likely to have adverse effects on the environment of the human environment and impede development. (*Principle 16*)

6. Scientific research and development in the context of environmental problems, both national and multinational, must be promoted in all countries, especially the developing countries. In this connection, the free flow of up-to-date scientific information and transfer of experience must be supported and assisted, to facilitate the solution of environmental problems; environmental technologies should be made available to developing countries on terms which would encourage their wide dissemination without constituting an economic

burden on the developing countries. (*Principle 20*)

Passage 1

Directions: *Read the following passage and choose the best answer for each of the following questions according to the information given in the passage.*

History of Environmental Policy Making

Public policies aimed at environmental protection date back to ancient times. The earliest sewers were constructed in Mohenjo-daro (Indus, or Harappan, civilization) and in Rome (ancient Roman civilization), which date back some 4,500 years and 2,700 years ago, respectively. Other civilizations implemented environmental laws. The city-states of ancient Greece created laws that governed forest harvesting some 2,300 years ago, and feudal European societies established hunting preserves, which limited game and timber harvesting to royalty, effectively preventing overexploitation, by 1000 CE. The city of Paris developed Europe's first large-scale sewer system during the 17th century. When the effects of industrialization and urbanization increased during the late 19th and early 20th centuries and threatened human health, governments developed additional rules and regulations for urban hygiene, sewage, sanitation, and housing, as well as the first laws devoted to protecting natural landscapes and wildlife (such as the creation of Yellowstone National Park as the world's first national park in 1872). Wealthy individuals and private foundations, such as the Sierra Club (founded 1892) and the National Audubon Society (founded 1905), also contributed to efforts to conserve natural resources and wildlife.

People became aware of the harmful effects of emissions and use of chemicals in industry and pesticides in agriculture during the 1950s and 1960s. The emergence of Minamata disease in 1956 in Japan, which resulted from mercury discharges from nearby chemical companies, and the publication of *Silent Spring* (1962) by American biologist Rachel Carson, which highlighted the dangers of pollution, led to a greater public awareness of environmental issues and to detailed systems of regulations in many industrialized countries. In those regulations, governments forbade the use of hazardous substances or prescribed maximum emission levels of specific substances to ensure a minimum environmental quality. Such regulative systems, like the Clean Water and Clean Air acts in the United States, succeeded in effectively addressing point sources (i.e., any discernable discrete location or piece of equipment that discharges pollution), such as industrial plants and utilities, where the cause-and-effect relationship between the actors causing the negative environmental effect could be clearly established.

Nevertheless, some environmental problems persisted, often because of the many nonpoint (diffuse) sources, such as exhaust from private automobiles and pesticide and fertilizer runoff from small farms, that contributed to air and water pollution. Individually, those small sources may not be harmful, but the accumulation of their pollution can exceed the regulative minimum norms for environmental quality. Also, the increasing complexity of chains of cause and effect has contributed to persistent problems. In the 1980s the effects of acid rain showed that the causes of environmental pollution could be separated geographically from its effects. Pollution problems of all types underscored the message that Earth's natural resources were being depleted and degraded.

From the late 1980s, sustainable development— (i.e., the fostering of economic growth while preserving the quality of the environment for future generations) —became a leading concept in environmental policy making. With nature and natural resources considered as economic drivers, environmental policy making was no longer the exclusive domain of government. Instead, private industry and nongovernmental organizations assumed greater responsibility for the environment. Also, the concept emphasized that individual people and their communities play a key role in the effective implementation of policies.

Over the years, a variety of guiding principles have been developed to help policy makers. Examples of such guiding principles, some of which have acquired a legal basis in some countries, are the "polluter pays" principle, which makes polluters liable for the costs of environmental damage, and the precautionary principle, which states that an activity is not allowed when there is a chance that the consequences are irreversible.

Such straightforward guiding principles do not work in all situations. For example, some environmental challenges, such as global warming, illuminate the need to view Earth as an ecosystem consisting of various subsystems, which, once disrupted, can lead to rapid changes that are beyond human control. Getting polluters to pay or the sudden adoption of the precautionary principle by all countries would not necessarily roll back the damage already imparted to the biosphere, though it would reduce future damage.

Since the early 1970s, environmental policies have made a shift from end-of-pipe solutions to prevention and control. Such solutions rely on the mitigation of negative effects. In addition, if a negative effect was unavoidable, it could be compensated for by investing in nature in other places than where the damage was caused, for example.

A third solution, which develops policies that focus on adapting the living environment to the change, is also possible. More specifically, measures that strengthen an ecosystem's ecological resilience (i.e., an ecosystem's ability to maintain its normal patterns of nutrient cycling and biomass production), combined with measures that emphasize prevention and mitigation, have been used. One such example is in Curitiba, Brazil, a city where some districts flood each year. The residents of flood-prone districts were relocated to higher and dryer places, and their former living areas were transformed into parks that could be flooded without disrupting city life.

1. According paragraph 1, which of the following statements is **NOT** true?_____
   A. The earliest environmental protection laws date back to Roman Empire.
   B. There were laws regarding forest felling in city-states of ancient Greece.

C. Hunting was limited to certain areas to avoid over-killing in feudal Europe.

D. The first laws on wildlife protection appeared at the turn of the 20th century.

2. The public awareness of environmental problems was greatly raised owing to _____.

A. the joint efforts made by many industrialized countries as well as individuals

B. governments' forbidding the use of chemicals in industry and agriculture

C. the appearance of Minamata disease and the publication of *Silent Spring*

D. implementation of some regulative systems in the United States

3. What can we learn from paragraph 3? _____

A. Many nonpoint sources may result in the persistence of some environmental problems.

B. Emissions from automobiles are more harmful than pesticide runoff from small farms.

C. Small nonpoint sources do not affect environmental quality because they may not be harmful.

D. All kinds of pollution problems show that natural resources on the earth are being severely wasted.

4. Which statement best summarizes the characteristic of environmental policy making from the late 1980s? _____

A. Preserving the quality of the environment is more important than developing economy.

B. Both individuals and communities play an important part in environmental policy making.

C. Governments assume the greatest responsibility in the domain of environmental policy making.

D. Sustainable development has become the major concern for environmental policy makers.

5. The adoption of the precautionary principle is likely to be helpful in _____.

A. bearing the consequences of environmental pollution

B. reducing future damage to the biosphere

C. avoid negative effects of challenges

D. preventing global warming

Passage 2

Directions: *In this section, there is a passage with twelve blanks. You are required to select one word for each blank from a list of choices given in a word bank following the passage. Read the passage through carefully before making your choices. Each choice in the bank is identified by a letter. You may not use any of the words in the bank more than once.*

A. bracketing	B. transboundary	C. forging	D. compromise
E. underpinnings	F. envisaged	G. instruments	H. apart
I. proper	J. intimately	K. stock	L. regarding

第 1 章　联合国人类环境会议宣言
Chapter 1　Declaration of the United Nations Conference on the Human Environment

The Stockholm and Rio Declarations are outputs of the first and second global environmental conferences, respectively, namely the United Nations Conference on the Human Environment in Stockholm, June 5 to 16, 1972, and the United Nations Conference on Environment and Development (UNCED) in Rio de Janeiro, June 3 to 14, 1992. Other policy or legal __1__ that emerged from these conferences, such as the *Action Plan for the Human Environment* at Stockholm and *Agenda 21* at Rio, are __2__ linked to the two declarations, conceptually as well as politically. However, the declarations, in their own right, represent signal achievements. Adopted twenty years __3__, they undeniably represent major milestones in the evolution of international environmental law, __4__ what has been called the "modern era" of international environmental law.

Stockholm represented a first taking __5__ of the global human impact on the environment, an attempt at __6__ a basic common outlook on how to address the challenge of preserving and enhancing the human environment. As a result, the *Stockholm Declaration* espouses mostly broad environmental policy goals and objectives rather than detailed normative positions. However, following Stockholm, global awareness of environmental issues increased dramatically, as did international environmental law-making __7__. At the same time, the focus of international environmental activism progressively expanded beyond __8__ and global commons issues to media-specific and cross-sectoral regulation and the synthesizing of economic and development considerations in environmental decision-making. By the time of the Rio Conference, therefore, the task for the international community became one of systematizing and restating existing normative expectations __9__ the environment, as well as of boldly positing the legal and political __10__ of sustainable development. In this vein, UNCED was expected to craft an "Earth Charter", a solemn declaration on legal rights and obligations bearing on environment and development, in the mold of the United Nations General Assembly's 1982 *World Charter for Nature* (General Assembly resolution 37/7). Although the __11__ text that emerged at Rio was not the lofty document originally __12__, the Rio Declaration, which reaffirms and builds upon the Stockholm Declaration, has nevertheless proved to be a major environmental legal landmark.

Further Studies and Post-Reading Discussion

Task 1
Directions: *Surf the Internet and find more information about* **Declaration of the United Nations Conference on the Human Environment**. *Work in groups and work out a report on one of the following topics.*

1. Purposes of *Declaration of the United Nations Conference on the Human Environment*.
2. Global objectives on preserving and improving human environment.
3. China's efforts in the protection and improvement of human environment.

Task 2

Directions: *Read the following sentences on Eco-Civilization and make a speech on your understanding of the eco-environmental conservation.*

绿水青山就是金山银山

绿水青山就是金山银山。这是重要的发展理念，也是推进现代化建设的重大原则。绿水青山就是金山银山，阐述了经济发展和生态环境保护的关系，揭示了保护生态环境就是保护生产力、改善生态环境就是发展生产力的道理，指明了实现发展和保护协同共生的新路径。 绿水青山既是自然财富、生态财富，又是社会财富、经济财富。保护生态环境就是保护自然价值和增值自然资本，就是保护经济社会发展潜力和后劲，使绿水青山持续发挥生态效益和经济社会效益。（摘自《习近平谈治国理政（第三卷）》）	Clear waters and green mountains are invaluable assets. This is an important concept of development and a major principle behind our modernization drive. It emphasizes the relationship between economic development and eco-environmental protection——this means preserving and developing productive forces. It points out a new way to coordinate development and protection. Clear waters and green mountains are not only natural and ecological wealth, but also social and economic wealth. Protecting the eco-environment means protecting nature's value and adding value to nature' capital, protecting the potential of economic and social development, and giving full play to the ecological, social and economic effects of nature. (Excerpt from *Xi Jinping: The Governance of China III*)

第 2 章　21 世纪议程

Chapter 2　Agenda 21

Background and Significance

《21 世纪议程》是 1992 年 6 月在巴西里约热内卢召开的联合国环境与发展大会通过的重要文件之一，旨在鼓励发展的同时保护环境。Agenda 有行动计划的意思，*Agenda 21* 就是面向 21 世纪的行动计划，是可持续发展所有领域全球行动的总体计划。该议程的签署是为确保地球未来的安全迈出的历史性的一步。

《21 世纪议程》中，各国政府提出了详细的行动蓝图，从而改变世界目前的非持续的经济增长模式，转向保护和更新经济增长和发展所依赖的环境资源的活动。行动领域包括保护大气层，阻止滥伐森林、水土流失和荒漠化，防止空气污染和水污染，预防渔业资源的枯竭，以及改进有毒废弃物的安全管理。

《21 世纪议程》还提出了引起环境压力的发展模式：发展中国家的贫穷和外债，非持续的生产和消费模式，人口压力和国际经济结构。行动计划提出了加强主要人群在实现可持续发展中所应起的作用，他们包括妇女、工会、农民、儿童和青年、土著人、科学界、当地政府、商界、工业界和非政府组织等。

为了全面支持在世界范围内落实《21 世纪议程》，联合国大会在 1992 年成立了可持续发展委员会。这个有 53 个成员的委员会监督并报告本议程和其他地球首脑会议的协议的执行情况，支持和鼓励政府、商界、工业界和其他非政府组织带来可持续发展所需要的社会和经济变化，帮助协调联合国内环境和发展的活动。

Text Study

Agenda 21 – Chapter 11
COMBATING DEFORESTATION
PROGRAMME AREAS

A. Sustaining the multiple roles and functions of all types of forests, forest lands and woodlands

Basis for action

11.1

There are major weaknesses in the policies, methods and mechanisms adopted to support and develop the multiple ecological, economic, social and cultural roles of trees, forests and forest lands. Many developed countries are confronted with the effects of air pollution and fire damage on their forests. More effective measures and approaches are often required at the national level to improve and harmonize policy formulation, planning and programming; legislative measures and instruments; development patterns; participation of the general public, especially women and indigenous people; involvement of youth; roles of the private sector, local organizations, non-governmental organizations and cooperatives; development of technical and multidisciplinary skills and quality of human resources; forestry extension and public education; research capability and support; administrative structures and mechanisms, including intersectoral coordination, decentralization and responsibility and incentive systems; and dissemination of information and public relations. This is especially important to ensure a rational and holistic approach to the sustainable and environmentally sound development of forests. The need for securing the multiple roles of forests and forest lands through adequate and appropriate institutional strengthening has been repeatedly emphasized in many of the reports, decisions and recommendations of FAO, ITTO, UNEP, the World Bank, IUCN and other organizations.

Objectives

11.2

The objectives of this programme area are as follows:

a. To strengthen forest-related national institutions, to enhance the scope and effectiveness of activities related to the management, conservation and sustainable development of forests, and to effectively ensure the sustainable utilization and production of forests' goods and services in both the developed and the developing countries; by the year 2000, to strengthen the capacities and capabilities of national institutions to enable them to acquire the necessary knowledge for the protection and conservation of forests, as well as to expand their scope and, correspondingly, enhance the effectiveness of programmes and activities related to the management and development of forests;

b. To strengthen and improve human, technical and professional skills, as well as expertise and capabilities to effectively formulate and implement policies, plans, programmes, research and projects on management, conservation and sustainable development of all types of forests and forest-based resources, and forest lands inclusive, as well as other areas from which forest benefits can be derived.

Activities

(a) Management-related activities

11.3

Governments at the appropriate level, with the support of regional, subregional and international organizations, should, where necessary, enhance institutional capability to promote the multiple roles and functions of all types of forests and vegetation inclusive of other related lands and forest-based resources in supporting sustainable development and environmental conservation in all sectors. This should be done, wherever possible and necessary, by strengthening and/or modifying the existing structures and arrangements, and by improving cooperation and coordination of their respective roles. Some of the major activities in this regard are as follows:

a. Rationalizing and strengthening administrative structures and mechanisms, including provision of adequate levels of staff and allocation of responsibilities, decentralization of decision-making, provision of infrastructural facilities and equipment, intersectoral coordination and an effective system of communication;

b. Promoting participation of the private sector, labour unions, rural cooperatives, local communities, indigenous people, youth, women, user groups and non-governmental organizations in forest-related activities, and access to information and training programmes within the national context;

c. Reviewing and, if necessary, revising measures and programmes relevant to all types of forests and vegetation, inclusive of other related lands and forest-based resources, and relating them to other land uses and development policies and legislation; promoting adequate legislation and other measures as a basis against uncontrolled conversion to other types of land uses;

d. Developing and implementing plans and programmes, including definition of national and, if necessary, regional and subregional goals, programmes and criteria for their implementation and subsequent improvement;

e. Establishing, developing and sustaining an effective system of forest extension and public education to ensure better awareness, appreciation and management of forests with regard to the multiple roles and values of trees, forests and forest lands;

f. Establishing and/or strengthening institutions for forest education and training, as well as forestry industries, for developing an adequate cadre of trained and skilled staff at the professional, technical and vocational levels, with emphasis on youth and women;

g. Establishing and strengthening capabilities for research related to the different aspects of forests and forest products, for example, on the sustainable management of forests, research on biodiversity, on the effects of air-borne pollutants, on traditional uses of forest resources by local populations and indigenous people, and on improving market returns and other non-market values from the management of forests.

(b) Data and information

11.4

Governments at the appropriate level, with the assistance and cooperation of international, regional, subregional and bilateral agencies, where relevant, should develop adequate databases and baseline information necessary for planning and programme evaluation. Some of the more specific activities include the following:

a. Collecting, compiling and regularly updating and distributing information on land classification and land use, including data on forest cover, areas suitable for afforestation, endangered species, ecological values, traditional/indigenous land use values, biomass and productivity, correlating demographic, socio-economic and forest resources information at the micro-and macro-levels, and undertaking periodic analyses of forest programmes;

 b. Establishing linkages with other data systems and sources relevant to supporting forest management, conservation and development, while further developing or reinforcing existing systems such as geographic information systems, as appropriate;

 c. Creating mechanisms to ensure public access to this information.

 (c) International and regional cooperation and coordination

11.5

Governments at the appropriate level and institutions should cooperate in the provision of expertise and other support and the promotion of international research efforts, in particular with a view to enhancing transfer of technology and specialized training and ensuring access to experiences and research results. There is need for strengthening coordination and improving the performance of existing forest-related international organizations in providing technical cooperation and support to interested countries for the management, conservation and sustainable development of forests.

Means of implementation

 (a) Financial and cost evaluation

11.6

The secretariat of the Conference has estimated the average total annual cost (1993—2000) of implementing the activities of this programme to be about $2.5 billion, including about $860 million from the international community on grant or concessional terms. These are indicative and order-of-magnitude estimates only and have not been reviewed by Governments. Actual costs and financial terms, including any that are non-concessional, will depend upon, inter alia, the specific strategies and programmes Governments decide upon for implementation.

 (b) Scientific and technological means

11.7

The planning, research and training activities specified will form the scientific and technological means for implementing the programme, as well as its output. The systems, methodology and know-how generated by the programme will help improve efficiency. Some of the specific steps involved should include:

 a. Analysing achievements, constraints and social issues for supporting programme formulation and implementation;

 b. Analysing research problems and research needs, research planning and implementation of specific research projects;

 c. Assessing needs for human resources, skill development and training;

 d. Developing, testing and applying appropriate methodologies/approaches in implementing forest programmes and plans.

(c) Human resource development

11.8

The specific components of forest education and training will effectively contribute to human resource development. These include:

a. Launching of graduate and post-graduate degree, specialization and research programmes;

b. Strengthening of pre-service, in-service and extension service training programmes at the technical and vocational levels, including training of trainers/teachers, and developing curriculum and teaching materials/methods;

c. Special training for staff of national forest-related organizations in aspects such as project formulation, evaluation and periodical evaluations.

(d) Capacity-building

11.9

This programme area is specifically concerned with capacity-building in the forest sector and all programme activities specified contribute to that end. In building new and strengthened capacities, full advantage should be taken of the existing systems and experience.

B. Enhancing the protection, sustainable management and conservation of all forests, and the greening of degraded areas, through forest rehabilitation, afforestation, reforestation and other rehabilitative means

Basis for action

11.10

Forests worldwide have been and are being threatened by uncontrolled degradation and conversion to other types of land uses, influenced by increasing human needs; agricultural expansion; and environmentally harmful mismanagement, including, for example, lack of adequate forest-fire control and anti-poaching measures, unsustainable commercial logging, overgrazing and unregulated browsing, harmful effects of airborne pollutants, economic incentives and other measures taken by other sectors of the economy. The impacts of loss and degradation of forests are in the form of soil erosion; loss of biological diversity, damage to wildlife habitats and degradation of watershed areas, deterioration of the quality of life and reduction of the options for development.

11.11

The present situation calls for urgent and consistent action for conserving and sustaining forest resources. The greening of suitable areas, in all its component activities, is an effective way of increasing public awareness and participation in protecting and managing forest resources. It should include the consideration of land use and tenure patterns and local needs and should spell out and clarify the specific objectives of the different types of greening activities.

Objectives

11.12

The objectives of this programme area are as follows:

a. To maintain existing forests through conservation and management, and sustain and expand areas under forest and tree cover, in appropriate areas of both developed and developing countries, through the conservation of natural forests, protection, forest rehabilitation, regeneration, afforestation, reforestation and tree planting, with a view to maintaining or restoring the ecological balance and expanding the contribution of forests to human needs and welfare;

b. To prepare and implement, as appropriate, national forestry action programmes and/or plans for the management, conservation and sustainable development of forests. These programmes and/or plans should be integrated with other land uses. In this context, country-driven national forestry action programmes and/or plans under the *Tropical Forestry Action Programme* are currently being implemented in more than 80 countries, with the support of the international community;

c. To ensure sustainable management and, where appropriate, conservation of existing and future forest resources;

d. To maintain and increase the ecological, biological, climatic, socio-cultural and economic contributions of forest resources;

e. To facilitate and support the effective implementation of the non-legally binding authoritative statement of principles for a global consensus on the management, conservation and sustainable development of all types of forests, adopted by the United Nations Conference on Environment and Development, and on the basis of the implementation of these principles to consider the need for and the feasibility of all kinds of appropriate internationally agreed arrangements to promote international cooperation on forest management, conservation and sustainable development of all types of forests, including afforestation, reforestation and rehabilitation.

Activities

(a) Management-related activities

11.13

Governments should recognize the importance of categorizing forests, within the framework of long-term forest conservation and management policies, into different forest types and setting up sustainable units in every region/watershed with a view to securing the conservation of forests. Governments, with the participation of the private sector, non-governmental organizations, local community groups, indigenous people, women, local government units and the public at large, should act to maintain and expand the existing vegetative cover wherever ecologically, socially and economically feasible, through technical cooperation and other forms of support. Major activities to be considered include:

a. Ensuring the sustainable management of all forest ecosystems and woodlands, through improved proper planning, management and timely implementation of silvicultural operations, including inventory and relevant research, as well as rehabilitation of degraded natural forests to restore productivity and environmental contributions, giving particular attention to human needs for economic and ecological services, wood-based energy, agroforestry, non-timber forest products and services, watershed and soil protection, wildlife management, and forest genetic resources;

 b. Establishing, expanding and managing, as appropriate to each national context, protected area systems, which includes systems of conservation units for their environmental, social and spiritual functions and values, including conservation of forests in representative ecological systems and landscapes, primary old-growth forests, conservation and management of wildlife, nomination of World Heritage Sites under the *World Heritage Convention*, as appropriate, conservation of genetic resources, involving in situ and ex situ measures and undertaking supportive measures to ensure sustainable utilization of biological resources and conservation of biological diversity and the traditional forest habitats of indigenous people, forest dwellers and local communities;

 c. Undertaking and promoting buffer and transition zone management;

 d. Carrying out revegetation in appropriate mountain areas, highlands, bare lands, degraded farm lands, arid and semi-arid lands and coastal areas for combating desertification and preventing erosion problems and for other protective functions and national programmes for rehabilitation of degraded lands, including community forestry, social forestry, agroforestry and silvipasture, while also taking into account the role of forests as national carbon reservoirs and sinks;

 e. Developing industrial and non-industrial planted forests in order to support and promote national ecologically sound afforestation and reforestation/regeneration programmes in suitable sites, including upgrading of existing planted forests of both industrial and non-industrial and commercial purpose to increase their contribution to human needs and to offset pressure on primary/old growth forests. Measures should be taken to promote and provide intermediate yields and to improve the rate of returns on investments in planted forests, through interplanting and underplanting valuable crops;

 f. Developing/strengthening a national and/or master plan for planted forests as a priority, indicating, inter alia, the location, scope and species, and specifying areas of existing planted forests requiring rehabilitation, taking into account the economic aspect for future planted forest development, giving emphasis to native species;

 g. Increasing the protection of forests from pollutants, fire, pests and diseases and other human-made interferences such as forest poaching, mining and unmitigated shifting cultivation, the uncontrolled introduction of exotic plant and animal species, as well as developing and accelerating research for a better understanding of problems relating to the management and regeneration of all types of forests; strengthening and/or establishing appropriate measures to assess and/or check inter-border movement of plants and related materials;

 h. Stimulating development of urban forestry for the greening of urban, peri-urban and rural human settlements for amenity, recreation and production purposes and for protecting trees and groves;

 i. Launching or improving opportunities for participation of all people, including youth, women, indigenous people and local communities in the formulation, development and implementation of forest-related programmes and other activities, taking due account of the local needs and cultural values;

j. Limiting and aiming to halt destructive shifting cultivation by addressing the underlying social and ecological causes.

(b) Data and information

11.14

Management-related activities should involve collection, compilation and analysis of data/information, including baseline surveys. Some of the specific activities include the following:

a. Carrying out surveys and developing and implementing land-use plans for appropriate greening/planting/afforestation/reforestation/forest rehabilitation;

b. Consolidating and updating land-use and forest inventory and management information for management and land-use planning of wood and non-wood resources, including data on shifting cultivation and other agents of forest destruction;

c. Consolidating information on genetic resources and related biotechnology, including surveys and studies, as necessary;

d. Carrying out surveys and research on local/indigenous knowledge of trees and forests and their uses to improve the planning and implementation of sustainable forest management;

e. Compiling and analysing research data on species/site interaction of species used in planted forests and assessing the potential impact on forests of climatic change, as well as effects of forests on climate, and initiating in-depth studies on the carbon cycle relating to different forest types to provide scientific advice and technical support;

f. Establishing linkages with other data/information sources that relate to sustainable management and use of forests and improving access to data and information;

g. Developing and intensifying research to improve knowledge and understanding of problems and natural mechanisms related to the management and rehabilitation of forests, including research on fauna and its interrelation with forests;

h. Consolidating information on forest conditions and site-influencing immission and emissions.

(c) International and regional cooperation and coordination

11.15

The greening of appropriate areas is a task of global importance and impact. The international and regional community should provide technical cooperation and other means for this programme area. Specific activities of an international nature, in support of national efforts, should include the following:

a. Increasing cooperative actions to reduce pollutants and trans-boundary impacts affecting the health of trees and forests and conservation of representative ecosystems;

b. Coordinating regional and subregional research on carbon sequestration, air pollution and other environmental issues;

c. Documenting and exchanging information/experience for the benefit of countries with similar problems and prospects;

d. Strengthening the coordination and improving the capacity and ability of intergovernmental organizations such as FAO, ITTO, UNEP and UNESCO to provide technical support for the management, conservation and sustainable development of forests, including support for the negotiation of the International Tropical Timber Agreement of 1983, due in 1992/1993.

Means of implementation

(a) Financial and cost evaluation

11.16

The secretariat of the Conference has estimated the average total annual cost (1993—2000) of implementing the activities of this programme to be about $10 billion, including about $3.7 billion from the international community on grant or concessional terms. These are indicative and order-of-magnitude estimates only and have not been reviewed by Governments. Actual costs and financial terms, including any that are non-concessional, will depend upon, inter alia, the specific strategies and programmes Governments decide upon for implementation.

(b) Scientific and technological means

11.17

Data analysis, planning, research, transfer/development of technology and/or training activities form an integral part of the programme activities, providing the scientific and technological means of implementation. National institutions should:

a. Develop feasibility studies and operational planning related to major forest activities;

b. Develop and apply environmentally sound technology relevant to the various activities listed;

c. Increase action related to genetic improvement and application of biotechnology for improving productivity and tolerance to environmental stress and including, for example, tree breeding, seed technology, seed procurement networks, germ-plasm banks, "in vitro" techniques, and in situ and ex situ conservation.

(c) Human resource development

11.18

Essential means for effectively implementing the activities include training and development of appropriate skills, working facilities and conditions, public motivation and awareness. Specific activities include:

a. Providing specialized training in planning, management, environmental conservation, biotechnology etc.;

b. Establishing demonstration areas to serve as models and training facilities;

c. Supporting local organizations, communities, non-governmental organizations and private land owners, in particular women, youth, farmers and indigenous people/shifting cultivators, through extension and provision of inputs and training.

(d) Capacity-building

11.19

National Governments, the private sector, local organizations/communities, indigenous people, labour unions and non-governmental organizations should develop capacities, duly supported by relevant international organizations, to implement the programme activities. Such capacities should be developed and strengthened in harmony with the programme activities. Capacity-building activities include policy and legal frameworks, national institution building, human resource development, development of research and technology, development of infrastructure, enhancement of public awareness etc.

C. Promoting efficient utilization and assessment to recover the full valuation of the goods and services provided by forests, forest lands and woodlands

Basis for action

11.20

The vast potential of forests and forest lands as a major resource for development is not yet fully realized. The improved management of forests can increase the production of goods and services and, in particular, the yield of wood and non-wood forest products, thus helping to generate additional employment and income, additional value through processing and trade of forest products, increased contribution to foreign exchange earnings, and increased return on investment. Forest resources, being renewable, can be sustainably managed in a manner that is compatible with environmental conservation. The implications of the harvesting of forest resources for the other values of the forest should be taken fully into consideration in the development of forest policies. It is also possible to increase the value of forests through non-damaging uses such as eco-tourism and the managed supply of genetic materials. Concerted action is needed in order to increase people's perception of the value of forests and of the benefits they provide. The survival of forests and their continued contribution to human welfare depends to a great extent on succeeding in this endeavour.

Objectives

11.21

The objectives of this programme area are as follows:

a. To improve recognition of the social, economic and ecological values of trees, forests and forest lands, including the consequences of the damage caused by the lack of forests; to promote methodologies with a view to incorporating social, economic and ecological values of trees, forests and forest lands into the national economic accounting systems; to ensure their sustainable management in a way that is consistent with land use, environmental considerations and development needs;

b. To promote efficient, rational and sustainable utilization of all types of forests and vegetation inclusive of other related lands and forest-based resources, through the development of efficient forest-based processing industries, value-adding secondary processing and trade in forest products, based on sustainably managed forest resources and in accordance with plans that integrate all wood and non-wood values of forests;

c. To promote more efficient and sustainable use of forests and trees for fuelwood and energy supplies;

d. To promote more comprehensive use and economic contributions of forest areas by incorporating eco-tourism into forest management and planning.

Activities

(a) Management-related activities

11.22

Governments, with the support of the private sector, scientific institutions, indigenous people, non-governmental organizations, cooperatives and entrepreneurs, where appropriate, should undertake the following activities, properly coordinated at the national level, with financial and technical cooperation from international organizations:

a. Carrying out detailed investment studies, supply-demand harmonization and environmental impact analysis to rationalize and improve trees and forest utilization and to develop and establish appropriate incentive schemes and regulatory measures, including tenurial arrangements, to provide a favourable investment climate and promote better management;

b. Formulating scientifically sound criteria and guidelines for the management, conservation and sustainable development of all types of forests;

c. Improving environmentally sound methods and practices of forest harvesting, which are ecologically sound and economically viable, including planning and management, improved use of equipment, storage and transportation to reduce and, if possible, maximize the use of waste and improve value of both wood and non-wood forest products;

d. Promoting the better use and development of natural forests and woodlands, including planted forests, wherever possible, through appropriate and environmentally sound and economically viable activities, including silvicultural practices and management of other plant and animal species;

e. Promoting and supporting the downstream processing of forest products to increase retained value and other benefits;

f. Promoting/popularizing non-wood forest products and other forms of forest resources, apart from fuelwood (e.g., medicinal plants, dyes, fibres, gums, resins, fodder, cultural products, rattan, bamboo) through programmes and social forestry/participatory forest activities, including research on their processing and uses;

g. Developing, expanding and/or improving the effectiveness and efficiency of forest-based processing industries, both wood and non-wood based, involving such aspects as efficient conversion technology and improved sustainable utilization of harvesting and process residues; promoting underutilized species in natural forests through research, demonstration and commercialization; promoting value-adding secondary processing for improved employment, income and retained value; and promoting/improving markets for, and trade in, forest products through relevant institutions, policies and facilities;

h. Promoting and supporting the management of wildlife, as well as eco-tourism, including farming, and encouraging and supporting the husbandry and cultivation of wild species, for improved rural income and employment, ensuring economic and social benefits without harmful ecological impacts;

i. Promoting appropriate small-scale forest-based enterprises for supporting rural development and local entrepreneurship;

j. Improving and promoting methodologies for a comprehensive assessment that will capture the full value of forests, with a view to including that value in the market-based pricing structure of wood and non-wood based products;

k. Harmonizing sustainable development of forests with national development needs and trade policies that are compatible with the ecologically sound use of forest resources, using, for example, the *ITTO Guidelines for Sustainable Management of Tropical Forests*;

l. Developing, adopting and strengthening national programmes for accounting the economic and non-economic value of forests.

(b) Data and information

11.23

The objectives and management-related activities presuppose data and information analysis, feasibility studies, market surveys and review of technological information. Some of the relevant activities include:

a. Undertaking analysis of supply and demand for forest products and services, to ensure efficiency in their utilization, wherever necessary;

b. Carrying out investment analysis and feasibility studies, including environmental impact assessment, for establishing forest-based processing enterprises;

c. Conducting research on the properties of currently underutilized species for their promotion and commercialization;

d. Supporting market surveys of forest products for trade promotion and intelligence;

e. Facilitating the provision of adequate technological information as a measure to promote better utilization of forest resources.

(c) International and regional cooperation and coordination

11.24

Cooperation and assistance of international organizations and the international community in technology transfer, specialization and promotion of fair terms of trade, without resorting to unilateral restrictions and/or bans on forest products contrary to GATT and other multilateral trade agreements, the application of appropriate market mechanisms and incentives will help in addressing global environmental concerns. Strengthening the coordination and performance of existing international organizations, in particular FAO, UNIDO, UNESCO, UNEP, ITC/UNCTAD/GATT, ITTO and ILO, for providing technical assistance and guidance in this programme area is another specific activity.

Means of implementation

(a) Financial and cost evaluation

11.25

The secretariat of the Conference has estimated the average total annual cost (1993—

2000) of implementing the activities of this programme to be about $18 billion, including about $880 million from the international community on grant or concessional terms. These are indicative and order-of-magnitude estimates only and have not been reviewed by Governments. Actual costs and financial terms, including any that are non-concessional, will depend upon, inter alia, the specific strategies and programmes Governments decide upon for implementation.

(b) Scientific and technological means

11.26

The programme activities presuppose major research efforts and studies, as well as improvement of technology. This should be coordinated by national Governments, in collaboration with and supported by relevant international organizations and institutions. Some of the specific components include:

a. Research on properties of wood and non-wood products and their uses, to promote improved utilization;

b. Development and application of environmentally sound and less-polluting technology for forest utilization;

c. Models and techniques of outlook analysis and development planning;

d. Scientific investigations on the development and utilization of non-timber forest products;

e. Appropriate methodologies to comprehensively assess the value of forests.

(c) Human resource development

11.27

The success and effectiveness of the programme area depends on the availability of skilled personnel. Specialized training is an important factor in this regard. New emphasis should be given to the incorporation of women. Human resource development for programme implementation, in quantitative and qualitative terms, should include:

a. Developing required specialized skills to implement the programme, including establishing special training facilities at all levels;

b. Introducing/strengthening refresher training courses, including fellowships and study tours, to update skills and technological know-how and improve productivity;

c. Strengthening capability for research, planning, economic analysis, periodical evaluations and evaluation, relevant to improved utilization of forest resources;

d. Promoting efficiency and capability of private and cooperative sectors through provision of facilities and incentives.

(d) Capacity-building

11.28

Capacity-building, including strengthening of existing capacity, is implicit in the programme activities. Improving administration, policy and plans, national institutions, human resources, research and scientific capabilities, technology development, and periodical evaluations and evaluation are important components of capacity-building.

D. Establishing and/or strengthening capacities for the planning, assessment and systematic observations of forests and related programmes, projects and activities, including commercial trade and processes

Basis for action

11.29

Assessment and systematic observations are essential components of long-term planning, for evaluating effects, quantitatively and qualitatively, and for rectifying inadequacies. This mechanism, however, is one of the often neglected aspects of forest resources, management, conservation and development. In many cases, even the basic information related to the area and type of forests, existing potential and volume of harvest is lacking. In many developing countries, there is a lack of structures and mechanisms to carry out these functions. There is an urgent need to rectify this situation for a better understanding of the role and importance of forests and to realistically plan for their effective conservation, management, regeneration, and sustainable development.

Objectives

11.30

The objectives of this programme area are as follows:

a. To strengthen or establish systems for the assessment and systematic observations of forests and forest lands with a view to assessing the impacts of programmes, projects and activities on the quality and extent of forest resources, land available for afforestation, and land tenure, and to integrate the systems in a continuing process of research and in-depth analysis, while ensuring necessary modifications and improvements for planning and decision-making. Specific emphasis should be given to the participation of rural people in these processes;

b. To provide economists, planners, decision makers and local communities with sound and adequate updated information on forests and forest land resources.

Activities

(a) Management-related activities

11.31

Governments and institutions, in collaboration, where necessary, with appropriate international agencies and organizations, universities and non-governmental organizations, should undertake assessments and systematic observations of forests and related programmes and processes with a view to their continuous improvement. This should be linked to related activities of research and management and, wherever possible, be built upon existing systems. Major activities to be considered are:

a. Assessing and carrying out systematic observations of the quantitative and qualitative situation and changes of forest cover and forest resources endowments, including land classification, land use and updates of its status, at the appropriate national level, and linking this activity, as appropriate, with planning as a basis for policy and programme formulation;

b. Establishing national assessment and systematic observation systems and evaluation of programmes and processes, including establishment of definitions, standards, norms and intercalibration methods, and the capability for initiating corrective actions as well as improving the formulation and implementation of programmes and projects;

　　c. Making estimates of impacts of activities affecting forestry developments and conservation proposals, in terms of key variables such as developmental goals, benefits and costs, contributions of forests to other sectors, community welfare, environmental conditions and biological diversity and their impacts at the local, regional and global levels, where appropriate, to assess the changing technological and financial needs of countries;

　　d. Developing national systems of forest resource assessment and valuation, including necessary research and data analysis, which account for, where possible, the full range of wood and non-wood forest products and services, and incorporating results in plans and strategies and, where feasible, in national systems of accounts and planning;

　　e. Establishing necessary intersectoral and programme linkages, including improved access to information, in order to support a holistic approach to planning and programming.

　　(b) Data and information

11.32

Reliable data and information are vital to this programme area. National Governments, in collaboration, where necessary, with relevant international organizations, should, as appropriate, undertake to improve data and information continuously and to ensure its exchange. Major activities to be considered are as follows:

　　a. Collecting, consolidating and exchanging existing information and establishing baseline information on aspects relevant to this programme area;

　　b. Harmonizing the methodologies for programmes involving data and information activities to ensure accuracy and consistency;

　　c. Undertaking special surveys on, for example, land capability and suitability for afforestation action;

　　d. Enhancing research support and improving access to and exchange of research results.

　　(c) International and regional cooperation and coordination

11.33

The international community should extend to the Governments concerned necessary technical and financial support for implementing this programme area, including consideration of the following activities:

　　a. Establishing conceptual framework and formulating acceptable criteria, norms and definitions for systematic observations and assessment of forest resources;

　　b. Establishing and strengthening national institutional coordination mechanisms for forest assessment and systematic observation activities;

　　c. Strengthening existing regional and global networks for the exchange of relevant information;

　　d. Strengthening the capacity and ability and improving the performance of existing international organizations, such as the Consultative Group on International Agricultural

Research (CGIAR), FAO, ITTO, UNEP, UNESCO and UNIDO, to provide technical support and guidance in this programme area.

Means of implementation

(a) Financial and cost evaluation

11.34

The secretariat of the Conference has estimated the average total annual cost (1993—2000) of implementing the activities of this programme to be about $750 million, including about $230 million from the international community on grant or concessional terms. These are indicative and order-of-magnitude estimates only and have not been reviewed by Governments. Actual costs and financial terms, including any that are non-concessional, will depend upon, inter alia, the specific strategies and programmes Governments decide upon for implementation.

11.35

Accelerating development consists of implementing the management-related and data/information activities cited above. Activities related to global environmental issues are those that will contribute to global information for assessing/evaluating/addressing environmental issues on a worldwide basis. Strengthening the capacity of international institutions consists of enhancing the technical staff and the executing capacity of several international organizations in order to meet the requirements of countries.

(b) Scientific and technological means

11.36

Assessment and systematic observation activities involve major research efforts, statistical modelling and technological innovation. These have been internalized into the management-related activities. The activities in turn will improve the technological and scientific content of assessment and periodical evaluations. Some of the specific scientific and technological components included under these activities are:

a. Developing technical, ecological and economic methods and models related to periodical evaluations and evaluation;

b. Developing data systems, data processing and statistical modelling;

c. Remote sensing and ground surveys;

d. Developing geographic information systems;

e. Assessing and improving technology.

11.37

These are to be linked and harmonized with similar activities and components in the other programme areas.

(c) Human resource development

11.38

The programme activities foresee the need and include provision for human resource development in terms of specialization (e.g., the use of remote-sensing, mapping and statistical modelling), training, technology transfer, fellowships and field demonstrations.

(d) Capacity-building

11.39

National Governments, in collaboration with appropriate international organizations and institutions, should develop the necessary capacity for implementing this programme area. This should be harmonized with capacity-building for other programme areas. Capacity-building should cover such aspects as policies, public administration, national-level institutions, human resource and skill development, research capability, technology development, information systems, programme evaluation, intersectoral coordination and international cooperation.

(e) Funding of international and regional cooperation

11.40

The secretariat of the Conference has estimated the average total annual cost (1993—2000) of implementing the activities of this programme to be about $750 million, including about $530 million from the international community on grant or concessional terms. These are indicative and order-of-magnitude estimates only and have not been reviewed by Governments. Actual costs and financial terms, including any that are non-concessional, will depend upon, inter alia, the specific strategies and programmes Governments decide upon for implementation.

Notes

1. GATT

《关税及贸易总协定》(*General Agreement on Tariffs and Trade*, GATT) 是一个政府间缔结的有关关税和贸易规则的多边国际协定，简称《关贸总协定》，是世界贸易组织（WTO）的前身。它的宗旨是通过削减关税和其他贸易壁垒，消除国际贸易中的差别待遇，促进国际贸易自由化，以充分利用世界资源，扩大商品的生产与流通。

《关贸总协定》于 1947 年 10 月 30 日在日内瓦签订，并于 1948 年 1 月 1 日开始临时适用。应当注意的是，由于未能达到规定的生效条件，作为多边国际协定的《关贸总协定》从未正式生效，而是一直通过《临时适用议定书》的形式产生临时适用的效力。

2. ITTO Guidelines for Sustainable Management of Tropical Forests

国际热带木材组织（International Tropical Timber Organization，ITTO）是《国际热带木材协定》的实施与管理机构，总部设在日本神户，其宗旨是为热带木材生产国和消费国之间提供一个合作和磋商的框架，促进和扩大热带木材的国际贸易以及改进热带木材市场的结构条件。国际热带木材组织在联合国环境发展大会之前为木材生产制定了一项可持续管理热带天然森林的国际参考标准，即 *Guidelines for the Sustainable Management of Natural Tropical Forests*，ITTO Policy Development Series No. 1.（国际热带木材组织政策制定系列第 1 号：热带天然林可持续管理准则）。1993 年，热带地区人工林的建立和管理准则、热带经济林生物多样性保护准则补充完善了国际热带木材组织的原则。各国都以国际热带木材组织参考标准为基础制定和执行国家准则的。

3. IUCN

世界自然保护联盟（International Union for Conservation of Nature, IUCN）于1948年在法国枫丹白露（Fontainebleau）成立，总部位于瑞士格朗，是世界上规模最大、历史最悠久的全球性非营利环保机构，也是自然环境保护与可持续发展领域唯一作为联合国大会永久观察员的国际组织。

IUCN是很独特的世界性联盟，是政府和非政府机构都能参加的少数几个国际组织之一，其会员组织分为主权国家和非营利机构，而各专家委员会则接受个人作为志愿成员加入。目前，有来自161个国家的200多个国家和政府机构会员、1 000多个非政府机构会员；超过16 000名学者个人会员加入专家委员会。IUCN目前在全球近50个国家设有办公室，有1 000多名雇员。

IUCN从20世纪80年代起就在中国开展工作。1996年，中华人民共和国外交部代表中国政府加入IUCN，中国成为国家会员。2003年成立中国联络处，2012年正式设立IUCN中国代表处。目前IUCN已有32个中国会员单位，其中包含香港4个。IUCN通过信息共享、国际交流、能力建设、地方示范项目等方式支持会员及合作伙伴开展工作，除为中国在重要的国际环境问题上与国际社会开展合作提供政策法规等技术支持外，还充分发挥联盟的全球性和区域性优势，开展了"大都市水源地可持续保护计划""中国保护地计划""生态系统生产总值（GEP）核算""未来红树林中国项目（MFF）"等项目。

4. *International Tropical Timber Agreement*

《国际热带木材协定》于1983年11月18日在日内瓦签订，1985年4月1日生效。中国于1986年7月2日交存加入书。1994年1月26日在日内瓦签订了新的《国际热带木材协定》，1997年1月1日零时生效。中国于1996年7月31日交存核准书，1997年1月1日生效。1994年的《国际热带木材协定》拥有51个成员，其中生产国为24个，都是发展中国家；消费国27个，除中国、埃及、尼泊尔、韩国、俄罗斯以外，其余都是发达市场经济国家。

《国际热带木材协定》是国际社会为保护热带森林生态系统，实现可持续利用和养护热带森林及其遗传资源而订立的国际法律文件。其宗旨是为生产和耗用热带木材的各国之间的合作和协商提供一个有效的纲领，促进国际热带木材贸易的扩展和多样化以及热带木材市场结构条件的改善，推广和支持研究与发展工作，以求改善森林管理和木材利用现状，鼓励制定旨在实现持久利用和养护热带森林及其遗传资源、保持有关区域生态平衡的各种国家政策。

5. *World Heritage Convention*

《世界遗产公约》是《保护世界文化和自然遗产公约》（*Convention Concerning the Protection of the World Cultural and Natural Heritage*）的简称，为联合国教育、科学及文化组织大会于1972年11月16日在巴黎签署，旨在为国际社会保护具有重大价值的文化和历史遗产建立一个长久性的有效制度。该公约主要规定了文化遗产和自然遗产的定义，以及文化和自然遗产的国家保护和国际保护的措施等条款。公约规定了各缔约国可自行确定本国领土内的文化和自然遗产，并向世界遗产委员会递交其遗产清单，由世界遗产大会审批。凡是被列入世界文化和自然遗产的地点，都由其所在国家依法严加保护。在全球范围内，共有187个国家或地区加入《世界遗产公约》，它是加入缔约国最多的公约之一。中国于 1985 年 12 月 12 日加入该公约。

Key Words and Phrases

1. afforestation	/ˌɒfrɪˈsteɪʃn/	n.	the conversion of bare or cultivated land into forest (originally for the purpose of hunting) 植树造林；绿化
2. buffer	/ˈbʌfə(r)/	n.	a thing or person that reduces a shock or protects sb./sth. against difficulties 缓冲；保护
3. concessional	/kənˈseʃənəl/	adj.	offered at a better rate than usual 优惠的；让步的
4. decentralization	/ˌdiːˌsentrəlaɪˈzeɪʃn/	n.	the act or process of giving some of the power of a central government, organization, etc. to smaller parts or organizations around the country 权力下放；分散
5. endangered	/ɪnˈdeɪndʒəd/	adj.	(of flora or fauna) in imminent danger of extinction 濒临灭绝的
6. expertise	/ˌekspɜːˈtiːz/	n.	expert knowledge or skill in a particular subject, activity or job 专门知识；专门技能；专长
7. husbandry	/ˈhʌzbəndri/	n.	the practice of cultivating the land or raising stock 饲养；农牧业
8. indigenous	/ɪnˈdɪdʒənəs/	adj.	originating where it is found 本土的；土著的
9. intercalibration	/ˌɪntəˌkælɪˈbreɪʃn/	n.	(technical) the mutual act of checking or adjusting by comparison with a standard 相互校准
10. interplanting	/ˌɪntəˈplɑːtɪŋ/	n.	the act of growing one type of crop alongside another type of crop 间种
11. intersectoral	/ˌɪntəˈsektərəl/	adj.	between or within different sectors 跨部门的
12. inventory	/ˈɪnvəntri/	n.	a list giving details of all the things in a place 存货清单
13. linkage	/ˈlɪŋkɪdʒ/	n.	the act of linking things; a link or system of links 连接；联系

14. magnitude	/'mægnɪtjuːd/	n.	the great size or importance of sth.; the degree to which sth. is large or important 巨大；重大；重要性
15. multidisciplinary	/ˌmʌltɪdɪsə'plɪnəri/	adj.	involving several different subjects of study （涉及）多门学科的
16. nomination	/ˌnɒmɪ'neɪʃn/	n.	the act of suggesting or choosing sb. as a candidate in an election, or for a job or an award; the fact of being suggested for this 提名；推荐
17. overgrazing	/ˌəʊvə'ɡreɪzɪŋ/	n.	intensive grazing for extended periods of time, or without sufficient recovery 过度放牧
18. rehabilitative	/riːhə'bɪlɪtətɪv/	adj.	designed to accomplish the original good condition 使复原的
19. residue	/'rezɪdjuː/	n.	matter that remains after something has been removed 加工后废料；残余
20. secretariat	/ˌsekrə'teəriət/	n.	the department of a large international or political organization which is responsible for running it, especially the office of a secretary general（大型国际组织、政治组织的）秘书处；书记处
21. silvicultural	/'sɪlvɪkʌltʃərəl/	adj.	related with the branch of forestry dealing with the development and care of forests 育林学的；造林术的
22. subregional	/sʌb'rɪdʒənəl/	adj.	of being ranked under a region 分区域的
23. tenurial	/ten'jʊriəl/	adj.	holding the right to property such as lands 土地保有的
24. underplanting	/ˌʌndə'plɑːntɪŋ/	n.	the act of planting between/under trees 套种；林下栽植
25. utilization	/ˌjuːtəlaɪ'zeɪʃn/	n.	the act of using; the state of having been made use of 利用；效用
26. vegetation	/ˌvedʒə'teɪʃ(ə)n/	n.	plants in general, especially the plants that are found in a particular area or environment （统称）植物；（尤指某地或环境的）植被，植物群落

27. air-borne pollutants 空气传导的污染物
28. germ-plasm banks 菌原生质库
29. in vitro /ɪn ˈviːtrəʊ/ 试管培育的；在生物体外进行的
30. labour union 工会
31. on grant or concessional terms 以赠予或减让条件方式
32. order of magnitude 量级估算；以重要等级划分
33. rural cooperatives 农村合作社

Exercises

Exercise 1 Reading Comprehension

Directions: *Read the excerpt of 11.1 through 11.4, and decide whether the following statements are true or false. Write T for true or F for false in the brackets in front of each statement.*

1. (   ) All the developed countries are confronted with the effects of air pollution and fire damage on their forests.

2. (   ) The need for securing the multiple roles of forests and forest lands through adequate and appropriate institutional strengthening has long been neglected in many of the reports, decisions and recommendations of FAO, ITTO, UNEP, the World Bank, IUCN and other organizations.

3. (   ) One of the objectives of this programme area is to strengthen forest-related national institutions, to enhance the scope and effectiveness of activities related to the management, conservation and sustainable development of forests, and to effectively ensure the sustainable utilization and production of forests' goods and services in both the developed and the developing countries.

4. (   ) Management-related activities should be done, wherever possible and necessary, by strengthening and/or modifying the existing structures and arrangements, and by improving cooperation and coordination of their respective roles.

5. (   ) We should establish, develop and sustain an effective system of forest extension and public education to ensure that the multiple roles and values of trees, forests and forest lands are appreciated and given top priority in the national economy.

6. (   ) Governments at the appropriate level, with the assistance and cooperation of international, regional, subregional and bilateral agencies, where relevant, should develop adequate databases and baseline information necessary for planning and programme evaluation.

Exercise 2 Skimming and Scanning

Directions: *Read the following passage excerpted from **Agenda 21**. At the end of the passage, there are six statements. Each statement contains information given in one of the paragraphs of the passage. Identify the paragraph from which the information is derived.*

Each paragraph is marked with a letter. You may choose a paragraph more than once. Answer the questions by writing the corresponding letter in the brackets in front of each statement.

11.13

Governments should recognize the importance of categorizing forests, within the framework of long-term forest conservation and management policies, into different forest types and setting up sustainable units in every region/watershed with a view to securing the conservation of forests. Governments, with the participation of the private sector, non-governmental organizations, local community groups, indigenous people, women, local government units and the public at large, should act to maintain and expand the existing vegetative cover wherever ecologically, socially and economically feasible, through technical cooperation and other forms of support. Major activities to be considered include:

A) Ensuring the sustainable management of all forest ecosystems and woodlands, through improved proper planning, management and timely implementation of silvicultural operations, including inventory and relevant research, as well as rehabilitation of degraded natural forests to restore productivity and environmental contributions, giving particular attention to human needs for economic and ecological services, wood-based energy, agroforestry, non-timber forest products and services, watershed and soil protection, wildlife management, and forest genetic resources;

B) Establishing, expanding and managing, as appropriate to each national context, protected area systems, which includes systems of conservation units for their environmental, social and spiritual functions and values, including conservation of forests in representative ecological systems and landscapes, primary old-growth forests, conservation and management of wildlife, nomination of World Heritage Sites under the *World Heritage Convention*, as appropriate, conservation of genetic resources, involving in situ and ex situ measures and undertaking supportive measures to ensure sustainable utilization of biological resources and conservation of biological diversity and the traditional forest habitats of indigenous people, forest dwellers and local communities;

C) Undertaking and promoting buffer and transition zone management;

D) Carrying out revegetation in appropriate mountain areas, highlands, bare lands, degraded farm lands, arid and semi-arid lands and coastal areas for combating desertification and preventing erosion problems and for other protective functions and national programmes for rehabilitation of degraded lands, including community forestry, social forestry, agroforestry and silvipasture, while also taking into account the role of forests as national carbon reservoirs and sinks;

E) Developing industrial and non-industrial planted forests in order to support and promote national ecologically sound afforestation and reforestation/regeneration programmes in suitable sites, including upgrading of existing planted forests of both industrial and non-industrial and commercial purpose to increase their contribution to human needs and to offset pressure on primary/old growth forests. Measures should be taken to promote and provide intermediate yields and to improve the rate of returns on investments in planted forests, through interplanting and underplanting valuable crops;

F) Developing/strengthening a national and/or master plan for planted forests as a priority, indicating, inter alia, the location, scope and species, and specifying areas of existing planted forests requiring rehabilitation, taking into account the economic aspect for future planted forest development, giving emphasis to native species;

G) Increasing the protection of forests from pollutants, fire, pests and diseases and other human-made interferences such as forest poaching, mining and unmitigated shifting cultivation, the uncontrolled introduction of exotic plant and animal species, as well as developing and accelerating research for a better understanding of problems relating to the management and regeneration of all types of forests; strengthening and/or establishing appropriate measures to assess and/or check inter-border movement of plants and related materials;

H) Stimulating development of urban forestry for the greening of urban, peri-urban and rural human settlements for amenity, recreation and production purposes and for protecting trees and groves;

I) Launching or improving opportunities for participation of all people, including youth, women, indigenous people and local communities in the formulation, development and implementation of forest-related programmes and other activities, taking due account of the local needs and cultural values;

J) Limiting and aiming to halt destructive shifting cultivation by addressing the underlying social and ecological causes.

1. (   ) Revegetation in appropriate and possible areas is vital for the prevention and control of desertification and soil erosion problems.

2. (   ) Industrial and non-industrial planted forests are to be developed for the purpose of supporting and promoting national ecologically sound afforestation and reforestation/regeneration programmes in suitable sites.

3. (   ) Forest poaching, mining and unmitigated shifting cultivation as well as the uncontrolled introduction of exotic plant and animal species are all examples of human-related problems.

4. (   ) To restore degraded natural forests to regain productivity and environmental contributions is part of the important steps of guaranteeing the sustainable management of all forest ecosystems and woodlands.

5. (   ) It is important that all people, including youth, women, indigenous people and local communities, be involved in the forest-related programmes.

6. (   ) Establishing, expanding and managing protected area systems, which includes systems of conservation units for their environmental, social and spiritual functions and values, should take each national background into consideration.

Exercise 3　Word Formation

Directions: *In this section, there are ten sentences from* ***Agenda 21****. You are required to complete these sentences with the proper form of the words given in blanks.*

1. The secretariat of the Conference has estimated the average total annual cost (1993—2000) of implementing the activities of this programme to be about $10 billion, including about $3.7 billion from the international community on grant or _____ terms. (concession)

2. The secretariat of the Conference has _____ the average total annual cost (1993—2000) of implementing the activities of this programme to be about $10 billion. (estimation)

3. National Governments, the private sector, local organizations/communities, indigenous people, labour unions and non-governmental organizations should develop capacities, duly supported by _____ international organizations, to implement the programme activities. (relevance)

4. Such capacities should be developed and strengthened in _____ with the programme activities. (harmonious)

5. Forest resources, being renewable, can be _____ managed in a manner that is compatible with environmental conservation. (sustain)

6. Developing, expanding and/or improving the effectiveness and efficiency of forest-based processing industries, both wood and non-wood based involve such aspects as efficient _____ technology and improved sustainable utilization of harvesting and process residues. (convert)

7. Concerted action is needed in order to increase people's _____ of the value of forests and of the benefits they provide. (perceive)

8. Governments, with the support of the private sector, scientific institutions, indigenous people, non-governmental organizations, _____ and entrepreneurs, where appropriate, should undertake the following activities. (cooperation)

9. Improving and promoting methodologies for a comprehensive _____ will capture the full value of forests. (assess)

10. Cooperation and assistance of international organizations and the international community in technology transfer, _____ and promotion of fair terms of trade, the application of appropriate market mechanisms and incentives will help in addressing global environmental concerns. (specialize)

Exercise 4 Translation

Section A

Directions: *Read Agenda 21, and complete the sentences by translating into English the Chinese given in blanks.*

1. There are major weaknesses in the policies, methods and mechanisms adopted to support and develop the multiple _____ (生态、经济、社会和文化作用) of trees, forests and forest lands.

2. Many developed countries are _____ (面临空气污染的影响) and fire damage on their forests.

3. This is especially important to ensure _____ (一种合理

和全盘考虑的办法) to the sustainable and environmentally sound development of forests.

4. The need for _____ (确保森林和林地的多种作用) through adequate and appropriate institutional strengthening has been repeatedly emphasized in many of the reports.

5. Governments at the appropriate level, with the support of regional, subregional and international organizations, should, where necessary, _____ (加强机构的能力) to promote the multiple roles and functions of all types of forests and vegetation.

6. These are indicative and order-of-magnitude estimates only and have not been reviewed by Governments. Actual costs and financial terms, _____ (包括任何非减让性条件), will depend upon, inter alia, the specific strategies and programmes Governments decide upon for implementation.

7. In this context, _____ (由国家主导的林业行动) programmes and/or plans under the *Tropical Forestry Action Programme* are currently being implemented in more than 80 countries, with the support of the international community.

8. Governments, with the participation of the private sector, _____ (非政府组织), local community groups, indigenous people, women, local government units and the public at large, should act to maintain and _____ (扩大现存植被覆盖面) wherever ecologically, socially and economically feasible, through technical cooperation and other forms of support.

9. Measures should be taken to promote and provide intermediate yields and _____ (提高投资的回报率) in planted forests, through interplanting and underplanting valuable crops.

10. Governments, with the support of the private sector, scientific institutions, indigenous people, non-governmental organizations, cooperatives and entrepreneurs, where appropriate, should undertake the following activities, _____ (经过国家层面适当协调), with financial and technical cooperation from international organizations.

Section B
Directions: *Translate the following sentences from English into Chinese.*

1. The success and effectiveness of the programme area depends on the availability of skilled personnel. Specialized training is an important factor in this regard. New emphasis should be given to the incorporation of women. (*11.27*)

2. Capacity-building, including strengthening of existing capacity, is implicit in the programme activities. Improving administration, policy and plans, national institutions, human resources, research and scientific capabilities, technology development, and periodical evaluations and evaluation are important components of capacity-building. (*11.28*)

3. To strengthen or establish systems for the assessment and systematic observations of forests and forest lands with a view to assessing the impacts of programmes, projects and activities on the quality and extent of forest resources, land available for afforestation, and

land tenure, and to integrate the systems in a continuing process of research and in-depth analysis, while ensuring necessary modifications and improvements for planning and decision-making. (*11.30.a*)

4. Making estimates of impacts of activities affecting forestry developments and conservation proposals, in terms of key variables such as developmental goals, benefits and costs, contributions of forests to other sectors, community welfare, environmental conditions and biological diversity and their impacts at the local, regional and global levels, where appropriate, to assess the changing technological and financial needs of countries. (*11.31. c*)

5. Assessment and systematic observation activities involve major research efforts, statistical modelling and technological innovation. These have been internalized into the management-related activities. The activities in turn will improve the technological and scientific content of assessment and periodical evaluations. (*11.36*)

6. The secretariat of the Conference has estimated the average total annual cost (1993—2000) of implementing the activities of this programme to be about $750 million, including about $530 million from the international community on grant or concessional terms. These are indicative and order-of-magnitude estimates only and have not been reviewed by Governments. Actual costs and financial terms, including any that are non-concessional, will depend upon, inter alia, the specific strategies and programmes Governments decide upon for implementation. (*11.40*)

Extensive Readings

Passage 1

Directions: *Read the following passage and choose the best answer for each of the following questions according to the information given in the passage.*

Review and challenges of policies of environmental protection and sustainable development in China

China is confronted with the dual task of developing its national economy and protecting its ecological environment. Since the 1980s, China's policies on environmental protection and sustainable development have experienced five changes: (1) progression from the adoption of environmental protection as a basic state policy to the adoption of sustainable development strategy; (2) changing focus from pollution control to ecological conservation equally; (3) shifting from end-of-pipe treatment to source control; (4) moving from point source treatment to regional environmental governance; (5) a turn away from administrative management-based approaches and towards a legal means and economic instruments-based approach.

Since 1992, China has set down sustainable development as a basic national strategy. However, environmental pollution and ecological degradation in China have continued to be serious problems and have inflicted great damage on the economy and quality of life. The beginning of the 21st century is a critical juncture for China's efforts towards sustaining rapid economic development, intensifying environmental protection efforts, and curbing ecological degradation. As the largest developing country, China's policies on environmental protection and sustainable development will be of primary importance not only for China, but also for the world. Realizing a completely well-off society by the year 2020 is seen as a crucial task by the Chinese government and an important goal for China's economic development in the new century; however, attaining it would require a four-fold increase over China's GDP in 2000. Therefore, speeding up economic development is a major mission during the next two decades and doing so will bring great challenges in controlling depletion of natural resources and environmental pollution. By taking a critical look at the development of Chinese environmental policy, we try to determine how best to coordinate the relationship between the environment and the economy in order to improve quality of life and the sustainability of China's resources and environment. Examples of important measures include: adjustment of economic structure, reform of energy policy, development of environmental industry, pollution prevention and ecological conservation, capacity building, and international cooperation and public participation.

China's environmental protection: challenges and countermeasures

The following suggestions are made to improve environmental quality in China: control population growth, develop ecofarming, protect biodiversity, explore new energy sources, practice safe and sustainable consumption, adopt an efficient strategy for use of natural resources, use industrial production to maintain sustainable development, strengthen management, guide urbanization, and increase international cooperation. The environmental problems are identified as the low use of industrial technology for preventing undesirable waste or emissions, increased sewage discharges in rural areas, water pollution, increased acid rain, increased pollution from solid waste and toxic chemicals, soil erosion and desertification, and agricultural pollution from chemical fertilizers.

Environmental pollution is the result of economic development. Economic development should, therefore, take responsibility for the solutions. Current funding for environmental protection is an inadequate 0.67% of the gross national product. Environmental management needs to be strengthened, given greater priority, and balanced with sustainable development. The entire population should be made aware of the importance of protecting the environment. Progress has been made in increasing cultivated forest preserves, in decreasing discharges of wastes into the water, and in decreasing noise pollution. 1,098 districts in 216 cities meet environmental noise control standards. The 1991 environmental protection staff consisted of 71,000 persons who were employed in 2,199 monitoring stations for soot control, 61 natural reserves, and 6,400 environmental pollution projects. A total investment of 1.74 billion RMB yuan was invested in environmental pollution control projects. Environmental quality has been maintained at 1980 standards, and further environmental pollution has been avoided. Atmospheric pollution in the third quarter of 1991 in large and medium-sized cities was 0.067 to 0.450 milligrams per cubic meter for total suspended particulates, 0.005 to 0.239 for sulphur dioxide concentration, and 0.012 to 0.139 for nitrogenous dioxides.

Chinese villages and their sustainable future: the European Union-China-Research Project "SUCCESS"

China has 800,000 villages—one person out of seven on the globe is living in a Chinese rural settlement. Yet the global discussion about the situation in China is currently characterized by a disproportionate focus on the development of towns and until now circumstances have generally been neglected in the rural areas, where 70% of the Chinese population is still living. Within the 5 years of the SUCCESS project research, this set of actual problems has been considered and analyzed under the principle of sustainability: "What to maintain?" "What to change?" were the overall research questions asked in the SUCCESS project; the researchers were looking for answers under a sustainability regime, respecting the need to raise the quality of life in the villages.

Several interweaving processes were used to achieve results: the inter-disciplinary research process between many areas of expertise, the trans-disciplinary process between the researchers and the Chinese villagers, and a negotiation process that made the connection between these two processes. The introduction describes the basic sustainability definition that was orienting the whole study. The innovation lies mostly in the methodology: the inter-disciplinary research co-operation related to practice and to involving the affected communities is needed to manage the significant and growing imbalances between urban and rural areas regarding their sustainability. In the transdisciplinary work, the project developed "village future sentences" that describe the local outcome of the research as one step towards better theoretical understanding of the mechanisms that could lead to a sustainable future, and they also managed to start sustainability processes in the case study sites. The integrated approach of the project helped generating future scenarios for these villages covering all aspects of their development, including urban design issues. Out of these scenarios, the villages developed small projects that could be implemented during the research period. This work made an important impact on community thinking within these villages.

However, it can also be seen as contributing to the dramatically changing development process in China, by finding a balance between traditional and contemporary approaches. In particular, the approach demonstrated a new, inter-disciplinary and trans-disciplinary negotiation processes whereby the local knowledge and the expert knowledge find common ground and outcomes. Only comprehensive concepts can contribute to an upgraded living standard, where living spaces and rural life should be recognized and esteemed in the future as a complement to urban lifestyles within the Chinese society. Innovative knowledge generation—such as the "systemic structure constellation" technique or the systems model approach—helped to bring out latent needs, hopes and potential of the villagers.

Besides the practical usage of these implemented projects, the process leading there showed the stakeholders their own fields of action. One major impact of these projects is the visibility of the results, which is crucial for villagers' awareness, their self-confidence and their experience with a successful participation in decision-making processes.

1. According to the first paragraph, which of the following is **NOT** the change that China

has experienced on its environmental policies since the 1980s?_____

    A. Changing focus from pollution control to ecological conservation equally.

    B. Moving from point source treatment to regional environmental governance.

    C. Shifting from rapid economic development to intensified protection efforts.

    D. Shifting from end-of-pipe treatment to source control.

2. According to the second paragraph, what significance will be achieved with China's policies on environmental protection and sustainable development?_____

    A. The global economic competition would witness the participation from China.

    B. China would demonstrate its own exemplary power for other developing countries in the world.

    C. It would be no longer an issue for us when considering natural resources depletion and environmental pollution.

    D. The general well-being from the China's society would be due to the implementation of these policies.

3. According to the fourth paragraph, what can be done in light of the current inadequate funding for environmental protection?_____

    A. Environmental management should be enhanced and prioritized.

    B. The public should be informed of the importance of ecosystem.

    C. Economic development needs to give way to environmental protection.

    D. The number of environmental protection staff needs to be reduced.

4. According to the fifth paragraph, why did the SUCCESS project focus on Chinese countryside?_____

    A. Because China has 800,000 villages, which shows their important roles.

    B. Because there is a lack of attention to the development of villages though most of the Chinese live there.

    C. Because Chinese villages have their unique characteristics.

    D. Because the global discussion about the situation in China has shifted from towns to villages.

5. Which of the following is **NOT** true about the approach adopted by the SUCCESS project?_____

    A. It's innovative and comprehensive.

    B. It's inter-disciplinary and trans-disciplinary.

    C. It features villagers' full participation in the research design.

    D. It strikes a balance between traditional and contemporary methods.

Passage 2

Directions: *In this section, there is a passage with twelve blanks. You are required to select one word for each blank from a list of choices given in a word bank following the passage. Read the passage through carefully before making your choices. Each choice in the bank is identified by a letter. You may not use any of the words in the bank more than once.*

A. mutual	B. section	C. binding	D. documents
E. minimize	F. explicitly	G. globally	H. expertise
I. vital	J. impacts	K. achieving	L. informal

NGOs and *Agenda 21*

Agenda 21 is an immense document of 40 chapters outlining an "action plan" for sustainable development, covering a wide range of specific natural resources and the role of different groups, as well as issues of social and economic development and implementation. It is one of the major __1__ that came out of United Nations' Rio Summit on Environment and Development in 1992. It is a comprehensive plan of action, recommended by UN Summit to be taken __2__, nationally, and locally by organizations of the UN System, Governments, and major groups in every area in which human __3__ the environment.

Thus, the influence of NGOs was even more relevant in the *Agenda 21* than in the climate convention due to its non-legally __4__ nature. *Agenda 21* is intended to set out an international program of action for __5__ sustainable development during the 21st century. *Agenda 21* is a report of more than 500 pages and comprising 40 chapters. Energy issues are __6__ spelled out in many chapters, although there is no single energy overview or focus reflecting political differences on this topic. In the __7__ on changing consumption patterns, NGOs provided language proposing to improve production efficiency to reduce energy and material use and to __8__ the generation of waste concepts reflected in the final text of *Agenda 21*. The text of *Agenda 21* is a particularly positive example of a collaborative approach and __9__ tolerance among governments and civil society, represented mostly by NGOs.

Agenda 21 also explicitly recognizes, in its chapter 27, the role of NGOs in implementing the agenda: Nongovernmental organizations play a __10__ role in shaping and implementing participatory democracy. Their credibility lies in the responsible and constructive role they play in society. Formal and __11__ organizations, as well as grassroots movements, should be recognized as partners in the implementation of the *Agenda 21*. Nongovernmental organizations possess well-established and diverse experience, __12__, and capacity in fields which will be of particular importance to the implementation and review of environmentally sound and socially responsible sustainable development.

Further Studies and Post-Reading Discussion

Task 1

Directions: *Surf the Internet and find more information about* ***Agenda 21****. Work in groups and work out a report on one of the following topics.*

1. Purposes of *Agenda 21*.
2. Current implementation of *Agenda 21* in the world and its paramount challenge in

recent years.

3. China's Agenda 21.

Task 2

Directions: *Read the following sentences on Eco-Civilization and make a speech on your understanding of the eco-environmental conservation.*

构建生态文明体系

2018年5月，全国生态环境保护大会总结并阐释了习近平生态文明思想。

习近平在大会上提出了新时代推进生态文明建设的主要原则，并首次提出加快构建生态文明体系，即以生态价值观念为准则的生态文化体系，以产业生态化和生态产业化为主体的生态经济体系，以改善生态环境质量为核心的目标责任体系，以治理体系和治理能力现代化为保障的生态文明制度体系，以生态系统良性循环和环境风险有效防控为重点的生态安全体系。

"五个体系"系统界定了生态文明体系的基本框架，指出了生态文明建设的思想保证、物质基础、制度保障以及责任和底线。"五个体系"不仅是建设美丽中国的行动指南，也为构建人类命运共同体贡献了生态文明理念和实践的"中国方案"。（摘自《中国关键词》生态文明篇）

One outcome of the National Conference on Eco-environmental Protection held in May 2018 was its summarization of Xi Jinping thought on eco-civilization.

Addressing the conference, Xi Jinping put forward the principles to apply in building an eco-civilization. For the first time, he raised the idea of developing a system for creating an eco-civilization. This means building a culture promoting eco-values, an economy highlighting eco-friendly industries and industrialization of environmental protection activities, a responsibility system for achieving the goal of improving the eco-environment, a system of institutions for building an eco-civilization supported by modernized governance, and a security system prioritizing well-functioning ecosystems and effective control of environmental risks.

This set a framework for building an eco-civilization based on a philosophy, material foundations, institutional support, clearly defined responsibilities, and an end-goal. It will guide the effort to build a beautiful China and offer a Chinese approach to eco-civilization in building a community with a shared future for humanity. (Excerpt from *Keywords to Understand China on Eco-Civilization*)

第 3 章 联合国千年宣言

Chapter 3 United Nations Millennium Declaration

Background and Significance

《联合国千年宣言》是一项旨在擘画人类千年发展框架的行动计划，于2000年9月6日至8日由联合国全体成员国在纽约联合国总部一致通过。

该宣言阐明了《二十一世纪国际议程》的价值、原则和目标，重申了联合国及《联合国宪章》在世界和平、繁荣和正义方面的不可或缺性，指出了所有的成员国都应该承担起维护人类尊严、平等、公平和保护儿童和弱势群体的责任。该宣言设定了消除极端贫穷和饥饿，普及小学教育，促进男女平等并赋予妇女权利，降低儿童死亡率，改善妇女保障，防治艾滋病毒/艾滋病、疟疾和其他疾病，确保环境可持续发展，全球合作促进发展8项目标和指标，统称为千年发展目标（Millennium Development Goals, 简称MDGs），成为国际社会最全面、权威、系统的发展目标体系。毫无疑问，《联合国千年宣言》在消除极端贫穷、维护环境可持续发展、保障人类福祉、推动全球发展等方面贡献卓越，它也是《2015年后发展议程》的重要依据。

Text Study

United Nations Millennium Declaration

The General Assembly

Adopts the following Declaration:

United Nations Millennium Declaration

Ⅰ. **Values and principles**

1. We, heads of State and Government, have gathered at United Nations Headquarters in New York from 6 to 8 September 2000, at the dawn of a new millennium, to reaffirm our faith in the Organization and its *Charter* as indispensable foundations of a more peaceful, prosperous and just world.

2. We recognize that, in addition to our separate responsibilities to our individual societies, we have a collective responsibility to uphold the principles of human dignity, equality and equity at the global level. As leaders we have a duty therefore to all the world's people, especially the most vulnerable and, in particular, the children of the world, to whom the future belongs.

3. We reaffirm our commitment to the purposes and principles of *the Charter of the United Nations*, which have proved timeless and universal. Indeed, their relevance and capacity to inspire have increased, as nations and peoples have become increasingly interconnected and interdependent.

4. We are determined to establish a just and lasting peace all over the world in accordance with the purposes and principles of *the Charter*. We rededicate ourselves to support all efforts to uphold the sovereign equality of all States, respect for their territorial integrity and political independence, resolution of disputes by peaceful means and in conformity with the principles of justice and international law, the right to self-determination of peoples which remain under colonial domination and foreign occupation, non-interference in the internal affairs of States, respect for human rights and fundamental freedoms, respect for the equal rights of all without distinction as to race, sex, language or religion and international cooperation in solving international problems of an economic, social, cultural or humanitarian character.

5. We believe that the central challenge we face today is to ensure that globalization becomes a positive force for all the world's people. For while globalization offers great opportunities, at present its benefits are very unevenly shared, while its costs are unevenly distributed. We recognize that developing countries and countries with economies in transition face special difficulties in responding to this central challenge. Thus, only through broad and sustained efforts to create a shared future, based upon our common humanity in all its diversity, can globalization be made fully inclusive and equitable. These efforts must include policies and measures, at the global level, which correspond to the needs of developing countries and economies in transition and are formulated and implemented with their effective participation.

6. We consider certain fundamental values to be essential to international relations in the twenty-first century. These include:

• **Freedom**. Men and women have the right to live their lives and raise their children in dignity, free from hunger and from the fear of violence, oppression or injustice. Democratic and participatory governance based on the will of the people best assures these rights.

• **Equality**. No individual and no nation must be denied the opportunity to benefit from development. The equal rights and opportunities of women and men must be assured.

• **Solidarity**. Global challenges must be managed in a way that distributes the costs and burdens fairly in accordance with basic principles of equity and social justice. Those who suffer or who benefit least deserve help from those who benefit most.

• **Tolerance**. Human beings must respect one other, in all their diversity of belief, culture and language. Differences within and between societies should be neither feared nor repressed, but cherished as a precious asset of humanity. A culture of peace and dialogue among all civilizations should be actively promoted.

• **Respect for nature**. Prudence must be shown in the management of all living species and natural resources, in accordance with the precepts of sustainable development. Only in this way can the immeasurable riches provided to us by nature be preserved and passed on to our descendants. The current unsustainable patterns of production and consumption must be changed in the interest of our future welfare and that of our descendants.

• **Shared responsibility**. Responsibility for managing worldwide economic and social development, as well as threats to international peace and security, must be shared among the nations of the world and should be exercised multilaterally. As the most universal and most representative organization in the world, the United Nations must play the central role.

7. In order to translate these shared values into actions, we have identified key objectives to which we assign special significance.

II. **Peace, security and disarmament**

8. We will spare no effort to free our peoples from the scourge of war, whether within or between States, which has claimed more than 5 million lives in the past decade. We will also seek to eliminate the dangers posed by weapons of mass destruction.

9. We resolve therefore:

• To strengthen respect for the rule of law in international as in national affairs and, in particular, to ensure compliance by Member States with the decisions of the International Court of Justice, in compliance with *the Charter of the United Nations*, in cases to which they are parties.

• To make the United Nations more effective in maintaining peace and security by giving it the resources and tools it needs for conflict prevention, peaceful resolution of disputes, peacekeeping, post-conflict peace-building and reconstruction. In this context, we take note of the report of the Panel on United Nations Peace Operations and request the General Assembly to consider its recommendations expeditiously.

• To strengthen cooperation between the United Nations and regional organizations, in accordance with the provisions of Chapter VIII of *the Charter*.

• To ensure the implementation, by States Parties, of treaties in areas such as arms control and disarmament and of *International Humanitarian Law and Human Rights Law*, and call upon all States to consider signing and ratifying *the Rome Statute of the International Criminal Court*.

• To take concerted action against international terrorism, and to accede as soon as possible to all the relevant international conventions.

• To redouble our efforts to implement our commitment to counter the world drug problem.

• To intensify our efforts to fight transnational crime in all its dimensions, including trafficking as well as smuggling in human beings and money laundering.

- To minimize the adverse effects of United Nations economic sanctions on innocent populations, to subject such sanctions regimes to regular reviews and to eliminate the adverse effects of sanctions on third parties.
- To strive for the elimination of weapons of mass destruction, particularly nuclear weapons, and to keep all options open for achieving this aim, including the possibility of convening an international conference to identify ways of eliminating nuclear dangers.
- To take concerted action to end illicit traffic in small arms and light weapons, especially by making arms transfers more transparent and supporting regional disarmament measures, taking account of all the recommendations of the forthcoming United Nations Conference on Illicit Trade in Small Arms and Light Weapons.
- To call on all States to consider acceding to the *Convention on the Prohibition of the Use, Stockpiling, Production and Transfer of Anti-personnel Mines and on Their Destruction*, as well as the amended mines protocol to the Convention on conventional weapons.

10. We urge Member States to observe *the Olympic Truce*, individually and collectively, now and in the future, and to support the International Olympic Committee in its efforts to promote peace and human understanding through sport and the Olympic Ideal.

III. Development and poverty eradication

11. We will spare no effort to free our fellow men, women and children from the abject and dehumanizing conditions of extreme poverty, to which more than a billion of them are currently subjected. We are committed to making the right to development a reality for everyone and to freeing the entire human race from want.

12. We resolve therefore to create an environment—at the national and global levels alike—which is conducive to development and to the elimination of poverty.

13. Success in meeting these objectives depends, inter alia, on good governance within each country. It also depends on good governance at the international level and on transparency in the financial, monetary and trading systems. We are committed to an open, equitable, rule-based, predictable and non-discriminatory multilateral trading and financial system.

14. We are concerned about the obstacles developing countries face in mobilizing the resources needed to finance their sustained development. We will therefore make every effort to ensure the success of the High-level International and Intergovernmental Event on Financing for Development, to be held in 2001.

15. We also undertake to address the special needs of the least developed countries. In this context, we welcome the Third United Nations Conference on the Least Developed Countries to be held in May 2001 and will endeavor to ensure its success. We call on the industrialized countries:

- To adopt, preferably by the time of that Conference, a policy of duty- and quota-free access for essentially all exports from the least developed countries;
- To implement the enhanced programme of debt relief for the heavily indebted poor countries without further delay and to agree to cancel all official bilateral debts of those

countries in return for their making demonstrable commitments to poverty reduction; and

• To grant more generous development assistance, especially to countries that are genuinely making an effort to apply their resources to poverty reduction.

16. We are also determined to deal comprehensively and effectively with the debt problems of low- and middle-income developing countries, through various national and international measures designed to make their debt sustainable in the long term.

17. We also resolve to address the special needs of small island developing States, by implementing *the Barbados Programme of Action* and the outcome of the twenty-second special session of the General Assembly rapidly and in full. We urge the international community to ensure that, in the development of a vulnerability index, the special needs of small island developing States are taken into account.

18. We recognize the special needs and problems of the landlocked developing countries, and urge both bilateral and multilateral donors to increase financial and technical assistance to this group of countries to meet their special development needs and to help them overcome the impediments of geography by improving their transit transport systems.

19. We resolve further:

• To halve, by the year 2015, the proportion of the world's people whose income is less than one dollar a day and the proportion of people who suffer from hunger and, by the same date, to halve the proportion of people who are unable to reach or to afford safe drinking water.

• To ensure that, by the same date, children everywhere, boys and girls alike, will be able to complete a full course of primary schooling and that girls and boys will have equal access to all levels of education.

• By the same date, to have reduced maternal mortality by three quarters, and under-five child mortality by two thirds, of their current rates.

• To have, by then, halted, and begun to reverse, the spread of HIV/AIDS, the scourge of malaria and other major diseases that afflict humanity.

• To provide special assistance to children orphaned by HIV/AIDS.

• By 2020, to have achieved a significant improvement in the lives of at least 100 million slum dwellers as proposed in the "Cities without Slums" initiative.

20. We also resolve:

• To promote gender equality and the empowerment of women as effective ways to combat poverty, hunger and disease and to stimulate development that is truly sustainable.

• To develop and implement strategies that give young people everywhere a real chance to find decent and productive work.

• To encourage the pharmaceutical industry to make essential drugs more widely available and affordable by all who need them in developing countries.

• To develop strong partnerships with the private sector and with civil society organizations in pursuit of development and poverty eradication.

• To ensure that the benefits of new technologies, especially information and communication technologies, in conformity with recommendations contained in *the ECOSOC 2000 Ministerial Declaration*, are available to all.

IV. **Protecting our common environment**

21. We must spare no effort to free all of humanity, and above all our children and grandchildren, from the threat of living on a planet irredeemably spoilt by human activities, and whose resources would no longer be sufficient for their needs.

22. We reaffirm our support for the principles of sustainable development, including those set out in *Agenda 21*, agreed upon at the United Nations Conference on Environment and Development.

23. We resolve therefore to adopt in all our environmental actions a new ethic of conservation and stewardship and, as first steps, we resolve:

• To make every effort to ensure the entry into force of the *Kyoto Protocol*, preferably by the tenth anniversary of the United Nations Conference on Environment and Development in 2002, and to embark on the required reduction in emissions of greenhouse gases.

• To intensify our collective efforts for the management, conservation and sustainable development of all types of forests.

• To press for the full implementation of the *Convention on Biological Diversity* and the *United Nations Convention to Combat Desertification in those Countries Experiencing Serious Drought and/or Desertification, Particularly in Africa*.

• To stop the unsustainable exploitation of water resources by developing water management strategies at the regional, national and local levels, which promote both equitable access and adequate supplies.

• To intensify cooperation to reduce the number and effects of natural and man-made disasters.

• To ensure free access to information on the human genome sequence.

V. **Human rights, democracy and good governance**

24. We will spare no effort to promote democracy and strengthen the rule of law, as well as respect for all internationally recognized human rights and fundamental freedoms, including the right to development.

25. We resolve therefore:

• To respect fully and uphold the *Universal Declaration of Human Rights*.

• To strive for the full protection and promotion in all our countries of civil, political, economic, social and cultural rights for all.

• To strengthen the capacity of all our countries to implement the principles and practices of democracy and respect for human rights, including minority rights.

• To combat all forms of violence against women and to implement the *Convention on the Elimination of All Forms of Discrimination against Women*.

• To take measures to ensure respect for and protection of the human rights of migrants, migrant workers and their families, to eliminate the increasing acts of racism and xenophobia in many societies and to promote greater harmony and tolerance in all societies.

• To work collectively for more inclusive political processes, allowing genuine participation by all citizens in all our countries.

• To ensure the freedom of the media to perform their essential role and the right of the public to have access to information.

VI. Protecting the vulnerable

26. We will spare no effort to ensure that children and all civilian populations that suffer disproportionately the consequences of natural disasters, genocide, armed conflicts and other humanitarian emergencies are given every assistance and protection so that they can resume normal life as soon as possible.

We resolve therefore:

• To expand and strengthen the protection of civilians in complex emergencies, in conformity with international humanitarian law.

• To strengthen international cooperation, including burden sharing in, and the coordination of humanitarian assistance to, countries hosting refugees and to help all refugees and displaced persons to return voluntarily to their homes, in safety and dignity and to be smoothly reintegrated into their societies.

• To encourage the ratification and full implementation of *the Convention on the Rights of the Child* and its optional protocols on the involvement of children in armed conflict and on the sale of children, child prostitution and child pornography.

VII. Meeting the special needs of Africa

27. We will support the consolidation of democracy in Africa and assist Africans in their struggle for lasting peace, poverty eradication and sustainable development, thereby bringing Africa into the mainstream of the world economy.

28. We resolve therefore:

• To give full support to the political and institutional structures of emerging democracies in Africa.

• To encourage and sustain regional and subregional mechanisms for preventing conflict and promoting political stability, and to ensure a reliable flow of resources for peacekeeping operations on the continent.

• To take special measures to address the challenges of poverty eradication and sustainable development in Africa, including debt cancellation, improved market access, enhanced Official Development Assistance and increased flows of Foreign Direct Investment, as well as transfers of technology.

• To help Africa build up its capacity to tackle the spread of the HIV/AIDS pandemic and other infectious diseases.

VIII. Strengthening the United Nations

29. We will spare no effort to make the United Nations a more effective instrument for pursuing all of these priorities: the fight for development for all the peoples of the world, the fight against poverty, ignorance and disease; the fight against injustice; the fight against violence, terror and crime; and the fight against the degradation and destruction of our

common home.

30. We resolve therefore:

• To reaffirm the central position of the General Assembly as the chief deliberative, policy-making and representative organ of the United Nations, and to enable it to play that role effectively.

• To intensify our efforts to achieve a comprehensive reform of the Security Council in all its aspects.

• To strengthen further the Economic and Social Council, building on its recent achievements, to help it fulfil the role ascribed to it in the *Charter*.

• To strengthen the International Court of Justice, in order to ensure justice and the rule of law in international affairs.

• To encourage regular consultations and coordination among the principal organs of the United Nations in pursuit of their functions.

• To ensure that the Organization is provided on a timely and predictable basis with the resources it needs to carry out its mandates.

• To urge the Secretariat to make the best use of those resources, in accordance with clear rules and procedures agreed by the General Assembly, in the interests of all Member States, by adopting the best management practices and technologies available and by concentrating on those tasks that reflect the agreed priorities of Member States.

• To promote adherence to *the Convention on the Safety of United Nations and Associated Personnel*.

• To ensure greater policy coherence and better cooperation between the United Nations, its agencies, the Bretton Woods Institutions and the World Trade Organization, as well as other multilateral bodies, with a view to achieving a fully coordinated approach to the problems of peace and development.

• To strengthen further cooperation between the United Nations and national parliaments through their world organization, the Inter-Parliamentary Union, in various fields, including peace and security, economic and social development, international law and human rights and democracy and gender issues.

• To give greater opportunities to the private sector, non-governmental organizations and civil society, in general, to contribute to the realization of the Organization's goals and programmes.

31. We request the General Assembly to review on a regular basis the progress made in implementing the provisions of this Declaration, and ask the Secretary-General to issue periodic reports for consideration by the General Assembly and as a basis for further action.

32. We solemnly reaffirm, on this historic occasion, that the United Nations is the indispensable common house of the entire human family, through which we will seek to realize our universal aspirations for peace, cooperation and development. We therefore pledge our unstinting support for these common objectives and our determination to achieve them.

8th plenary meeting
8 September 2000

Notes

1. Bretton Woods Institutions

布雷顿森林机构,指布雷顿森林体系建立的两大国际金融机构,即国际货币基金组织(IMF)和世界银行(World Bank)。前者负责向成员国提供短期资金借贷,目的为保障货币体系的稳定;后者提供中长期信贷来促进成员国经济复苏。

1944年7月1日,44个国家或政府的经济特使在美国新罕布什尔州的布雷顿森林召开了联合国货币金融会议(简称布雷顿森林会议),商讨战后国际货币体系问题。经过3周的讨论,会议通过了两个重要决议,即《国际货币基金协定》和《国际复兴开发银行协定》,确立了以美元为中心的国际货币体系,即布雷顿森林体系。该体系以美元和黄金为基础的金汇兑本位制,其实质是建立一种以美元为中心的国际货币体系,基本内容包括美元与黄金挂钩、国际货币基金会员国的货币与美元保持固定汇率(实行固定汇率制度)。布雷顿森林货币体系的运转与美元的信誉和地位密切相关。

布雷顿森林体系的建立,促进了战后资本主义世界经济的恢复和发展。因美元危机与美国经济危机的频繁爆发,以及制度本身不可解脱的矛盾性,该体系于1971年8月15日被尼克松政府宣告结束。1973年2月美元进一步贬值,世界各主要货币由于受投机商冲击被迫实行浮动汇率制,至此布雷顿森林体系完全崩溃。布雷顿森林体系崩溃以后,国际货币基金组织和世界银行作为重要的国际组织仍得以继续存在至今,并发挥重要的国际作用。

2. Convention on Biological Diversity

《生物多样性公约》是一项旨在保护濒临灭绝的植物和动物,最大限度地保护地球上的多种多样的生物资源的国际公约。该公约于1992年6月1日在联合国环境规划署发起的政府间谈判委员会第七次会议上通过,1992年6月5日由签约国在巴西里约热内卢举行的联合国环境与发展大会上签署,1993年12月29日正式生效。常设秘书处设在加拿大的蒙特利尔。

联合国《生物多样性公约》缔约方大会是全球履行该公约的最高决策机构,自1994年起,每两年数千名来自不同国家的代表齐聚缔约方大会,讨论如何保护生物多样性。中国于1992年6月11日签署该公约,于2021年10月主办了第十五次缔约方大会。

3. Olympic Truce

在古代希腊,每当举行奥运会期间,各交战城邦之间必须停战,以便各地的运动员参加比赛,并让各地观众前往观看,违反者将被禁止参加以后的奥运会,这项休战协议被称为"奥林匹克休战"。在1992年巴塞罗那奥运会时,挪威提议恢复奥林匹克休战的精神,在奥运开幕前一周起至闭幕后一周的期间内停战。这一旨在推动国际和平的提议获得国际奥委会的认可,并得到与会169个国家的认同。南斯拉夫内战因1994年冬季奥林匹克运动会而暂时休战,是现代第一个因奥林匹克休战原则而暂时休战的战争。之后,每届夏季奥运及冬季奥运的主办国都会要求参与国签署休战协议,虽然仍有在奥运期间未休战的情况。

联合国在1993年10月25日通过的第48/11号决议,呼吁会员国在奥运开幕前7天到闭幕后7天遵守奥林匹克休战原则。在《联合国千年宣言》中也再度在"和平、安全与裁军"一节中提及:"我们促请会员国从今以后都要遵守奥林匹克休战原则,并支持国际奥林匹克委员会努力通过体育和奥林匹克理想促进和平及人与人之间的相互谅解。"

4. Rome Statute of the International Criminal Court

《国际刑事法院罗马规约》（简称《规约》），是旨在保护国际人权、打击国际犯罪的刑事法律。1998 年 6 月，联合国大会在罗马召开了为期 5 周的外交大会，旨在拟定并通过一个有关建立国际刑事法院的公约。7 月 17 日，国际刑事法院全权代表外交会议在 21 个国家弃权的情况下，以 120 票赞成、7 票反对获得通过，按照《规约》第 126 条规定，该《规约》于 2002 年 7 月 1 日生效。

5. United Nations Convention to Combat Desertification in those Countries Experiencing Serious Drought and/or Desertification, Particularly in Africa

《联合国关于在发生严重干旱和/或荒漠化的国家特别是在非洲防治荒漠化的公约》，简称《联合国防治荒漠化公约》（United Nations Convention to Combat Desertification，UNCCD，以下简称《公约》），1994 年 6 月 17 日在法国巴黎通过，1996 年 12 月 26 日正式生效，是联合国环境与发展大会框架下的三大环境公约之一；截至 2005 年 4 月 26 日，该《公约》共有 191 个缔约方。《公约》的核心目标是由各国政府共同制定国家级、次区域级和区域级行动方案，并与捐助方、地方社区和非政府组织合作，以对抗应对荒漠化的挑战。履约资金匮乏、资金运作机制不畅一直是困扰《公约》发展的难题。2019 年 2 月，美国国家航天局研究结果表明，全球从 2000 年到 2017 年新增的绿化面积中，约 1/4 来自中国，中国贡献比例居全球首位。2019 年 2 月 26 日，《联合国防治荒漠化公约》第十三次缔约方大会第二次主席团会议在我国贵阳市举行。

Key Words and Phrases

1. abject	/'æbdʒekt/	adj.	[usually before noun] (formal) terrible and without hope 悲惨绝望的；凄惨的
2. accede	/ək'siːd/	vi.	(~ to sth.) (formal) to agree to a request, proposal, etc. 答应；同意
3. concerted	/kən'sɜːtɪd/	adj.	involving the joint activity of two or more 联合的；同心协力的
4. conducive	/kən'djuːsɪv/	adj.	(~ to sth.) making the other thing likely to happen 有助（于…）的；有益（于…）的
5. deliberative	/dɪ'lɪb(ə)rətɪv/	adj.	having the power or the right to make important decisions（机构或程序）审议的，有审议权的，起审议作用的
6. disarmament	/dɪs'ɑːməmənt/	n.	the act of reducing the number of weapons, especially nuclear weapons, that a country has 裁军；裁减军备（尤指核武器）

7. empowerment	/ɪm'paʊəmənt/	n.	the process of giving them power and status in a particular situation 授权
8. equity	/'ekwəti/	n.	(formal) a situation in which everyone is treated equally 公平；公正
9. expeditiously	/ˌekspə'dɪʃəsli/	adv.	quick and efficient 迅速地，有效地
10. genocide	/'dʒenəsaɪd/	n.	the murder of a whole race or group of people 种族灭绝；大屠杀
11. illicit	/ɪ'lɪsɪt/	adj.	contrary to or forbidden by law or the social customs of a country 违法的；不正当的
12. impediment	/ɪm'pedɪmənt/	n.	sth. that delays or stops the progress of sth. 妨碍；阻碍；障碍
13. implementation	/ˌɪmplɪmen'teɪʃ(ə)n/	n.	the act of accomplishing some aim or executing some order 实施；执行；贯彻
14. infectious	/ɪn'fekʃəs/	adj.	easily spread to others 传染性的，感染的
15. irredeemably	/ˌɪrɪ'diːməbli/	adv.	that cannot be corrected, improved or saved 不能改正地；无法补救地；不能挽救地
16. landlocked	/'lændlɒkt/	adj.	almost or completely surrounded by land 几乎（或完全）被陆地包围的；陆地环绕的；内陆的
17. millennium	/mɪ'leniəm/	n.	the time when one period of 1000 years ends and another begins 一千年；千年期
18. mortality	/mɔː'tæləti/	n.	the number of deaths in a particular situation or period of time 死亡数量；死亡率
19. pandemic	/pæn'demɪk/	n.	a disease that spreads over a whole country or the whole world（全国或全球性）流行病；大流行病
		adj.	(a disease) existing everywhere（疾病）大流行的；普遍的
20. pharmaceutical	/ˌfɑːmə'suːtɪk(ə)l/	adj.	connected with making and selling drugs and medicines 制药的

21. pornography	/pɔːˈnɒgrəfi/	n.	(disapproving) books, videos, etc. that describe or show naked people and sexual acts in order to make people feel sexually excited, especially in a way that many other people find offensive 淫秽作品；色情书刊（或音像制品等）
22. protocol	/ˈprəʊtəkɒl/	n.	a written record of a treaty or agreement that has been made by two or more countries 议定书；协议
23. ratify	/ˈrætɪfaɪ/	vt.	to make an agreement officially valid by voting for or signing it 正式批准；使正式生效
24. relevance	/ˈreləvəns/	n.	importance or significance of sth.; the relation of something to the matter at hand 重要性，意义；相关性
25. sanction	/ˈsæŋkʃn/	n.	a severe course of action which is intended to make people obey instructions, customs, or laws 处罚；制裁；惩罚
26. scourge	/skɜːdʒ/	n.	sth. that causes a lot of trouble or suffering to a group of people 灾难；祸害
27. solidarity	/ˌsɒlɪˈdærəti/	n.	support for each other or for another group, especially in political or international affairs 团结一致，相互支持
28. stewardship	/ˈstjuːədʃɪp/	n.	(formal) the act of taking care of or managing sth., for example property, an organization, money or valuable objects 管理；看管；组织工作
29. smuggling	/ˈsmʌglɪŋ/	n.	the crime of taking, sending or bringing goods secretly and illegally into or out of a country 走私（罪）
30. truce	/truːs/	n.	an agreement between two people or groups of people to stop fighting or quarrelling for a short time; the period of time that this lasts 停战协定；休

战；停战期

31. unevenly	/ʌnˈiːvnli/	adv.	in an unequal or partial manner; in an irregular way 不公平的；不均衡的；不规则地；不平坦地
32. unstinting	/ʌnˈstɪntɪŋ/	adj.	very generous 慷慨的；大方的
33. xenophobia	/ˌzenəˈfəʊbiə/	n.	(disapproving) a strong feeling of dislike or fear of people from other countries 仇外，惧外
34. inter alia	/ˌɪntər ˈeɪliə/	adv.	(from Latin, formal) among other things 除了其他事物之外

35. quota-free　　免配额
36. in compliance with　　依照，遵照
37. in conformity with　　与…相符，符合；遵照
38. under colonial domination　　在殖民统治下

Exercises

Exercise 1　Reading Comprehension

Directions: Read the **United Nations Millennium Declaration**, and decide whether the following statements are true or false. Write T for true or F for false in the brackets in front of each statement.

1. (　　) Globalization offers great opportunities for and brings fair benefits to all the nations, so the central challenge facing us today is to ensure that globalization becomes a positive force for the entire world's people.

2. (　　) Because no less than 5 million people died in wars in the past decade, it's necessary for us to take every measure to avoid disasters of war.

3. (　　) Our planet has been irreversibly damaged by human activities and can no longer provide sufficient resources for people's needs.

4. (　　) The media should enjoy freedom to report news and the public has the right to learn about information.

5. (　　) Refugees prove to be a burden to any country hosting refugees; therefore, they can be sent back to their homes against their will.

6. (　　) Being an effective organ, the United Nations can make policies related to issues such as development, poverty reduction, injustice, violence and degradation of the earth without taking account of the interests of all Member States.

Exercise 2　Skimming and Scanning

Directions: Read the following passage excerpted from the **United Nations Millennium Declaration**. At the end of the passage, there are six statements. Each statement contains information given in one of the paragraphs of the passage. Identify the paragraph from which *the information is derived. Each paragraph is marked with a letter. You may choose a paragraph more than once. Answer the questions by writing the corresponding letter in the brackets in front of each statement.*

III. Development and Poverty Eradication

A) We will spare no effort to free our fellow men, women and children from the abject and dehumanizing conditions of extreme poverty, to which more than a billion of them are currently subjected. We are committed to making the right to development a reality for everyone and to freeing the entire human race from want.

B) We resolve therefore to create an environment—at the national and global levels alike—which is conducive to development and to the elimination of poverty.

C) Success in meeting these objectives depends, inter alia, on good governance within each country. It also depends on good governance at the international level and on transparency in the financial, monetary and trading systems. We are committed to an open, equitable, rule-based, predictable and non-discriminatory multilateral trading and financial system.

D) We are concerned about the obstacles developing countries face in mobilizing the resources needed to finance their sustained development. We will therefore make every effort to ensure the success of the High-level International and Intergovernmental Event on Financing for Development, to be held in 2001.

E) We also undertake to address the special needs of the least developed countries. In this context, we welcome the Third United Nations Conference on the Least Developed Countries to be held in May 2001 and will endeavour to ensure its success. We call on the industrialized countries:

• To adopt, preferably by the time of that Conference, a policy of duty- and quota-free access for essentially all exports from the least developed countries.

• To implement the enhanced programme of debt relief for the heavily indebted poor countries without further delay and to agree to cancel all official bilateral debts of those countries in return for their making demonstrable commitments to poverty reduction.

• To grant more generous development assistance, especially to countries that are genuinely making an effort to apply their resources to poverty reduction.

F) We are also determined to deal comprehensively and effectively with the debt problems of low- and middle-income developing countries, through various national and international measures designed to make their debt sustainable in the long term.

G) We resolve further:

• To halve, by the year 2015, the proportion of the world's people whose income is less than one dollar a day and the proportion of people who suffer from hunger and, by the same date, to halve the proportion of people who are unable to reach or to afford safe drinking water.

• To ensure that, by the same date, children everywhere, boys and girls alike, will be able to complete a full course of primary schooling and that girls and boys will have equal access to all levels of education.

H) We also resolve to address the special needs of small island developing States, by implementing *the Barbados Programme of Action* and the outcome of the twenty-second special session of the General Assembly rapidly and in full. We urge the international community to ensure that, in the development of a vulnerability index, the special needs of small island developing States are taken into account.

I) We also resolve:

• To promote gender equality and the empowerment of women as effective ways to combat poverty, hunger and disease and to stimulate development that is truly sustainable.

• To develop and implement strategies that give young people everywhere a real chance to find decent and productive work.

• To encourage the pharmaceutical industry to make essential drugs more widely available and affordable by all who need them in developing countries.

1. (   ) Efforts should be made to reduce the population whose income is less than one dollar a day by 50%.

2. (   ) Good governance is essential to fulfill the development and poverty eradication objectives at home and abroad.

3. (   ) The industrialized countries are required to provide generous development assistance to poor countries that do their best to utilize their resources for the purpose of reducing poverty.

4. (   ) It is common that some essential drugs are in shortage and unaffordable in developing countries.

5. (   ) The programme of debt relief should be further promoted and the debts of heavily indebted poor countries should be written off if these countries are dedicated to poverty elimination.

6. (   ) The right to development is not a reality for all yet and there are more than a billion people still suffering from extreme poverty.

Exercise 3   Word Formation

Directions: *In this section, there are ten sentences from the* **United Nations Millennium Declaration**. *You are required to complete these sentences with the proper form of the words given in blanks.*

1. Thus, only through broad and sustained efforts to create a shared future, based upon our common humanity in all its diversity, can globalization be made fully _____ _____and equitable. (include)

2. We resolve therefore to create an environment—at the national and global levels alike—which is _____ to development and to the elimination of poverty. (conduce)

3. The current _____ patterns of production and consumption must be changed in the interest of our future welfare and our descendants. (sustain)

4. Responsibility for managing worldwide economic and social development, as well as

threats to international peace and security, must be shared among the nations of the world and should be exercised _____. (lateral)

5. We will spare no effort to free our fellow men, women and children from the abject and _____ conditions of extreme poverty, to which more than a billion of them are currently subjected. (human)

6. It also depends on good governance at the international level and on transparency in the financial, _____ and trading systems. (money)

7. We resolve therefore to adopt in all our environmental actions a new ethic of conservation and _____. (steward)

8. We will spare no effort to ensure that children and all civilian populations that suffer _____ the consequences of natural disasters, genocide, armed conflicts and other humanitarian emergencies are given every assistance and protection so that they can resume normal life as soon as possible. (proportion)

9. We will support the _____ of democracy in Africa and assist Africans in their struggle for lasting peace, poverty eradication and sustainable development, thereby bringing Africa into the mainstream of the world economy. (solid)

10. We resolved therefore to take special measures to address the challenges of poverty eradication and sustainable development in Africa, including debt _____, improved market access, enhanced Official Development Assistance and increased flows of Foreign Direct Investment, as well as transfers of technology. (cancel)

Exercise 4 Translation

Section A

Directions: *Read the* **United Nations Millennium Declaration**, *and complete the sentences by translating into English the Chinese given in blanks.*

1. We resolve to strengthen respect for the rule of law in international as in national affairs and, in particular, to ensure _____ (会员国遵守国际法院的判决), in compliance with *the Charter* of the United Nations, in cases to which they are parties.

2. In this context, we take note of the report of the _____ (联合国和平行动问题小组) and request the General Assembly to consider its recommendations expeditiously.

3. We resolve to ensure the implementation, by States Parties, of treaties in areas such as arms control and disarmament and of _____ (国际人道主义法和人权法), and call upon all States to consider signing and ratifying the *Rome Statute of the International Criminal Court*.

4. We resolve to intensify our efforts to fight transnational crime in all its dimensions, including _____ (贩卖和偷运人口以及洗钱行为).

5. We shall strive for _____ (消除大规模毁灭性武器), particularly nuclear weapons, and to keep all options open for achieving this aim, including the possibility of convening an international conference to identify ways of eliminating

nuclear dangers.

6. We urge Member States to observe the Olympic Truce, individually and collectively, now and in the future, and to support the International Olympic Committee in its efforts to _____ (通过体育和奥林匹克理想促进和平及人与人之间的互相理解).

7. We resolve to make every effort to ensure the entry into force of the Kyoto Protocol, preferably by the tenth anniversary of the United Nations Conference on Environment and Development in 2002, and to embark on _____ (按规定减少温室气体的排放).

8. We resolve to press for _____ (全面执行《生物多样性公约》) and the *Convention to Combat Desertification in those Countries Experiencing Serious Drought and/or Desertification, Particularly in Africa.*

9. We resolve to ensure greater policy coherence and better cooperation between the United Nations, its agencies, the Bretton Woods Institutions and the World Trade Organization, as well as other multilateral bodies, with a view to _____ (对和平与发展问题采取全面协调的对策).

10. We resolve to give greater opportunities to _____ (私营部门、非政府组织和广大民间社会), in general, to contribute to the realization of the Organization's goals and programmes.

Section B

Directions: *Translate the following sentences from English into Chinese.*

1. We are determined to establish a just and lasting peace all over the world in accordance with the purposes and principles of *the Charter*. We rededicate ourselves to support all efforts to uphold the sovereign equality of all States, respect for their territorial integrity and political independence, resolution of disputes by peaceful means and in conformity with the principles of justice and international law, the right to self-determination of peoples which remain under colonial domination and foreign occupation, non-interference in the internal affairs of States, respect for human rights and fundamental freedoms, respect for the equal rights of all without distinction as to race, sex, language or religion and international cooperation in solving international problems of an economic, social, cultural or humanitarian character. (*United Nations Millennium Declaration 4*)

2. We believe that the central challenge we face today is to ensure that globalization becomes a positive force for all the world's people. For while globalization offers great opportunities, at present its benefits are very unevenly shared, while its costs are unevenly distributed. We recognize that developing countries and countries with economies in transition face special difficulties in responding to this central challenge. Thus, only through broad and sustained efforts to create a shared future, based upon our common humanity in all its diversity, can globalization be made fully inclusive and equitable. These efforts must include policies and measures, at the global level, which correspond to the needs of developing countries and economies in transition and are formulated and implemented with their effective

participation. (*United Nations Millennium Declaration 5*)

3. Respect for nature. Prudence must be shown in the management of all living species and natural resources, in accordance with the precepts of sustainable development. Only in this way can the immeasurable riches provided to us by nature be preserved and passed on to our descendants. The current unsustainable patterns of production and consumption must be changed in the interest of our future welfare and that of our descendants. (*United Nations Millennium Declaration 6*)

4. We also resolve to address the special needs of small island developing States, by implementing the *Barbados Programme of Action* and the outcome of the twenty-second special session of the General Assembly rapidly and in full. We urge the international community to ensure that, in the development of a vulnerability index, the special needs of small island developing States are taken into account. (*United Nations Millennium Declaration 17*)

5. We recognize the special needs and problems of the landlocked developing countries, and urge both bilateral and multilateral donors to increase financial and technical assistance to this group of countries to meet their special development needs and to help them overcome the impediments of geography by improving their transit transport systems. (*United Nations Millennium Declaration 18*)

6. We solemnly reaffirm, on this historic occasion, that the United Nations is the indispensable common house of the entire human family, through which we will seek to realize our universal aspirations for peace, cooperation and development. We therefore pledge our unstinting support for these common objectives and our determination to achieve them. (*United Nations Millennium Declaration 32*)

Extensive Readings

Passage 1

Directions: *Read the following passage and choose the best answer for each of the following questions according to the information given in the passage.*

Goal 7 of the Millennium Development Goals Report (2005):
Ensure environmental sustainability

Target: *Integrate the principles of sustainable development into country policies and programmes and reverse the loss of environmental resources.*

Environmental sustainability means using natural resources wisely and protecting the complex ecosystems on which our survival depends. But sustainability will not be achieved with current patterns of resource consumption and use. Land is becoming degraded at an alarming rate. Plant and animal species are being lost in record numbers. The climate is changing, bringing with it threats of rising sea levels and worsening droughts and floods. Fisheries and other marine resources are being overexploited. The rural poor are most immediately affected because their day-to-day subsistence and livelihoods more often depend on the natural resources around them. Though the exodus to urban areas has reduced pressure on rural lands, it has increased the number of people living in unsafe and overcrowded urban slums. In both urban and rural areas, billions of people lack safe drinking water and basic sanitation. Overcoming these and other environmental problems will require greater attention to the plight of the poor and an unprecedented level of global cooperation. Action to halt further destruction of the ozone layer shows that progress is possible when the political will is there.

Most countries have committed to the principles of sustainable development and to incorporating them into their national policies and strategies. They have also agreed to the implementation of relevant international accords. But good intentions have not resulted in sufficient progress to reverse the loss of our environmental resources.

Forests are disappearing fastest in the poorest regions

Forests cover one third of the earth's surface and constitute one of the richest ecosystems. They provide for many people's everyday needs, including food, fuel, building materials and clean water. Yet, in the last decade alone, 940,000 square kilometers of forests—an area the size of Venezuela—were converted into farmland, logged or lost to other uses. Efforts to combat deforestation are ongoing. Sustainable forest management practices are reducing pressure on the land and improving the livelihoods of communities living in and around forests. Still, it is a race against time.

More areas are protected, but loss of species and habitats continues

Some 19 million square kilometers—over 13% of the earth's land surface—have been designated as protected areas. This represents an increase of 15% since 1994. The expansion of protected areas is encouraging, but their management does not always meet conservation goals. Moreover, marine environments are highly underrepresented, with less than 1% of marine ecosystems protected. Loss of habitats and biological diversity continues, with more than 10,000 species considered to be under threat.

Progress is being made in improving energy efficiency, but more is needed

Progress is being made in improving energy efficiency and access to clean technology and fuels. But the transfer of these new technologies to developing countries, where energy needs are skyrocketing, is not proceeding fast enough. Despite improved efficiency, total energy use continues to rise. In developing countries, the lack of clean fuels has a direct impact on rural households which depend on wood, dung, crop residues and charcoal for cooking and heating. Indoor air pollution caused by these fuels is estimated to cause more than 1.6 million deaths per year, mostly among women and children.

Rich countries produce the most greenhouse gases

The consumption of fossil fuels, including oil, coal and natural gas, results in carbon dioxide emissions that are contributing to the gradual warming of the planet. The expected repercussions of climate change—including rising sea waters, more frequent and intense storms, the extinction of species, worsening droughts and crop failures —will affect every nation on earth. With total emissions continuing to grow, the majority of industrialized countries have adopted the *Kyoto Protocol*, the first global effort to control emissions.

Ozone-depleting substances have been drastically reduced

The ozone layer in the stratosphere absorbs ultraviolet radiation, which has been associated with rising levels of skin cancer and other harmful effects on living species. Through unprecedented global cooperation, use of chlorofluorocarbons, the most widespread ozone-depleting substances, has been reduced to one tenth of 1990 levels. This remarkable accomplishment shows that progress on the environment can be achieved with strong political will and with consensus on the problem and on how to solve it. Though damage to the ozone layer is already evident, recovery is expected within the next 50 years.

Access to safe drinking water has improved worldwide

During the 1990s, access to improved drinking water sources increased substantially. However, over a billion people have yet to benefit, with lowest coverage in rural areas and urban slums. Much slower progress has been made globally in improving sanitation. An estimated 2.6 billion people—representing half the developing world— lack toilets and other forms of improved sanitation.

The proportion of population using safe sources of drinking water in the developing world rose from 71% in 1990 to 79% in 2002. The most impressive gains were made in Southern Asia. This jump was fueled primarily by increased coverage in India, home to over 1 billion people. The good news—gains in all regions since 1990—is counterbalanced by the fact that 1.1 billion people were still using water from unimproved sources in 2002. In sub-Saharan Africa, where 42% of the population is still unserved, the obstacles to progress, which include conflict, political instability and low priority assigned to investments in water and sanitation, are especially daunting given high population growth rates.

Half the developing world lacks improved sanitation

Sanitation coverage in the developing world rose from 34% in 1990 to 49% in 2002. If present trends continue, however, close to 2.4 billion people worldwide will still be without improved sanitation in 2015, that is, almost as many as there are today. The sanitation target can be met only with a dramatic increase in investment in services.

The growth in the number of slum-dwellers is outpacing urban improvement

Together, Southern Asia, Eastern Asia and sub-Saharan Africa account for more than two thirds of people living in slums. In most regions, countries are making efforts to provide alternatives to the formation of slums. But because of the rapid expansion in urban populations, the number of slum-dwellers is increasing in all developing regions, except Northern Africa.

1. According to the first paragraph, what might be the most effective way to ensure environmental sustainability?_____

  A. Evaluating the rate at which land is degrading.

  B. Changing current modes of resources consumption and use.

  C. Recording the number of extinct plant and animal species.

  D. Bringing worsening droughts and floods under control.

2. Why are the rural poor people easily affected by environmental problems?_____
   A. Because they do not know how to reduce pressure on rural lands.
   B. Because they lack safe drinking water and basic sanitation.
   C. Because they rely heavily on the natural resources for a living.
   D. Because they are too poor to overcome environmental problems.
3. What might be one of the solutions to preventing further loss of species and habitats?_____
   A. To improve the management of the protected areas and increase the protected marine area.
   B. To make high conservation goals and encourage an increase of protected area by 15 % each year.
   C. To conduct a thorough exploration of marine environments and make good use of them.
   D. To designate over 13% of the earth's land surface as protected areas and reduce species under threat.
4. Which of the following statements is true about the ozone layer?_____
   A. Damage to the ozone layer is considered as irreversible, and, therefore, should be avoided.
   B. The ozone layer contains some harmful substances causing skin cancer and loss of species.
   C. Reducing some widespread ozone-depleting substances is not useful for protecting the ozone layer.
   D. Political will and agreements play important roles in halting the depletion of the ozone layer.
5. In sub-Saharan Africa, what is **NOT** one of the reasons that results in having no access for large population to safe drinking water?_____
   A. Unstable political situations.
   B. High population growth rates.
   C. Fights among ethnic or tribal groups.
   D. Low budget spent on water.

Passage 2

Directions: *In this section, there is a passage with twelve blanks. You are required to select one word for each blank from a list of choices given in a word bank following the passage. Read the passage through carefully before making your choices. Each choice in the bank is identified by a letter. You may not use any of the words in the bank more than once.*

A. denied	B. partnership	C. persist	D. respective
E. range	F. consistency	G. adoption	H. crucial
I. tragically	J. comprehensive	K. unstable	L. entitled

第3章 联合国千年宣言
Chapter 3 United Nations Millennium Declaration

Foreword of the Millennium Development Goals Report (2005)

The adoption of the Millennium Development Goals, drawn from the *United Nations Millennium Declaration*, was a seminal event in the history of the United Nations. It constituted an unprecedented promise by world leaders to address, as a single package, peace, security, development, human rights and fundamental freedoms. As I said in my March 2005 report __1__ "In larger freedom: towards development, security and human rights for all", to which the present report is a complement: "We will not enjoy development without security, we will not enjoy security without development, and we will not enjoy either without respect for human rights. Unless all these causes are advanced, none will succeed." The eight Millennium Development Goals __2__ from halving (减半；均摊) extreme poverty to halting the spread of HIV/AIDS and providing universal primary education—all by the target date of 2015. They form a blueprint agreed by all the world's countries and all the world's leading development institutions—a set of simple but powerful objectives that every man and woman in the street, from New York to Nairobi to New Delhi, can easily support and understand. Since their __3__, the Goals have galvanized (激励；刺激) unprecedented efforts to meet the needs of the world's poorest.

Why are the Millennium Development Goals so different? There are four reasons. First, the Millennium Development Goals are people-centered, time-bound and measurable. Second, they are based on a global __4__, stressing the responsibilities of developing countries for getting their own house in order, and of developed countries for supporting those efforts. Third, they have unprecedented political support, embraced at the highest levels by developed and developing countries, civil society and major development institutions alike. Fourth, they are achievable.

The year 2005 is __5__ in our work to achieve the Goals. In September—5 years after they adopted the Millennium Declaration and 10 years before the Goals fall due — world leaders will meet at the United Nations in New York to assess how far their pledges have been fulfilled, and to decide on what further steps are needed. In many ways, the task this year will be much tougher than it was in 2000. Instead of setting targets, this time leaders must decide how to achieve them. THIS PROGRESS REPORT is the most __6__ accounting to date on how far we have come, and how far we have to go, in each of the world's regions. It reflects a collaborative effort among a large number of agencies and organizations within and outside the United Nations system. All have provided the most up-to-date data possible in their areas of responsibility, helping thereby to achieve clarity and __7__ in the report. Above all, the report shows us how much progress has been made in some areas, and how large an effort is needed to meet the Millennium Development Goals in others. If current trends __8__, there is a risk that many of the poorest countries will not be able to meet many of them. Considering how far we have come, such a failure would mark a __9__ missed opportunity. This report shows that we have the means at hand to ensure that nearly every country can make good on the promises of the Goals. Our challenge is to deploy (利用) those means.

As I said in my March report: "Let us be clear about the costs of missing this opportunity: millions of lives that could have been saved will be lost; many freedoms that could have been secured will be __10__; and we shall inhabit a more dangerous and __11__ world." I commend this report as a key resource in preparing for the September summit, which must be a time of decision. The analysis and information contained here can help citizens, civic organizations, Governments, parliaments and international bodies to play their ___12___ roles in making the Millennium Development Goals a reality. (Kofi a. Annan Secretary-General)

Further Studies and Post-Reading Discussion

Task 1
Directions: Surf the Internet and find more information about **the United Nations Millennium Declaration**. Work in groups and work out a report on one of the following topics.
1. Purposes of *United Nations Millennium Declaration*.
2. Progress made through the United Nations Millennium Development Goals.
3. China's practice and achievements in poverty eradication.

Task 2
Directions: Read the following sentences on Eco-Civilization and make a speech on your understanding of the eco-environmental conservation.

On Green Development

绿色发展，就其要义来讲，是要解决好人与自然和谐共生问题。绿色发展的真谛是"取之有度，用之有节"，就是要调整人的行为、纠正人的错误行为，从一味地利用自然、征服自然、改造自然向尊重自然、顺应自然、保护自然转变，改变长期以来"大量生产、大量消耗、大量排放"的生产模式和消费模式，把经济活动、人的行为限制在自然资源和生态环境能够承受的限度内，给自然生态留下休养生息的时间和空间。

中国作为全球生态文明建设的重要参与者、贡献者、引领者，主张加快构筑尊崇自然、绿色发展的生态体系，共建清洁美丽世界。中国愿与世界各国共同呵护好地球家园，同筑生态文明之基，同走绿色发展之路。（摘自《中国关键词》生态文明篇）

The purpose of green development, fundamentally speaking, is to achieve harmony between humanity and nature. Green development means taking from nature at the proper time and to the proper extent. To this end, we as human beings have to regulate our behavior, correct our wrongdoings, and abandon the old idea of utilizing, conquering and transforming nature in favor of respecting, accommodating and protecting nature. We must change the traditional model characterized by massive production, massive consumption and massive emissions, and keep economic and human activities within the carrying capacity of natural resources and the eco-environment, thus leaving time and space for nature to recuperate.

As an important participant, contributor and trailblazer in developing a global eco-civilization, China advocates jointly building a clean, beautiful world that respects nature and favors green development. (Excerpt from *Keywords to Understand China on Eco-Civilization*)

第 4 章　联合国森林文书

Chapter 4　United Nations Forest Instrument

Background and Significance

《联合国森林文书》，又称《森林文书》，最早为《关于所有类型森林的无法律约束力文书》(*The Non-Legally Binding Instrument on all Types of Forests*，NLBI)，是 2007 年 5 月在联合国森林论坛（The United Nations Forum on Forests，UNFF）通过的一项关于可持续森林管理的非法律约束力的文件。联合国森林论坛建立于 2000 年，隶属联合国经济及社会理事会（Economic and Social Council），是依照联合国 2000 年第 2000/35 号决议设立的。联合国森林论坛成员包括联合国所有成员国和常驻观察员、联合国森林论坛秘书处、森林合作伙伴关系、区域组织和进程以及主要集团。该论坛的设立突显出森林保护的全球重要性。而作为全球森林可持续发展中的一个里程碑，《森林文书》的通过证明了促进全球可持续森林管理迈出新的一步。《森林文书》为各国提供了促进可持续森林管理的框架，对加强国际合作，协调与森林有关的各种政策，减少森林砍伐、防止森林退化、促进可持续发展具有重大意义。

2015 年，第 11 届联合国森林论坛将《关于所有类型森林的无法律约束力文书》更名为《联合国森林文书》，进一步阐明了在国际和国家各级加强森林治理、技术和体制能力、政策和法律框架、森林部门投资和利益相关方参与的一系列政策和措施。

Text Study

United Nations Forest Instrument[①]

Member States,

Recognizing that forests and trees outside forests provide multiple economic, social and

---

① See General Assembly Resolution 70/199, the non-legally binding instrument on all types of forests was renamed as the United Nations forest instrument.

environmental benefits, and emphasizing that sustainable forest management contributes significantly to sustainable development and poverty eradication;

Recalling the *Non-legally Binding Authoritative Statement of Principles for a Global Consensus on Management, Conservation and Sustainable Development of All Types of Forests (Forest Principles)*;[①] chapter 11 of *Agenda 21*;[②] the proposals for action of the Intergovernmental Panel on Forests/Intergovernmental Forum on Forests; resolutions and decisions of the United Nations Forum on Forests; the *Johannesburg Declaration on Sustainable Development* and the *Plan of Implementation of the World Summit on Sustainable Development*;[③] the *Monterrey Consensus of the International Conference on Financing for Development*;[④] the internationally agreed development goals, including the Millennium Development Goals; the 2005 World Summit Outcome;[⑤] and existing international legally binding instruments relevant to forests;

Welcoming the accomplishments of the international arrangement on forests since its inception by the Economic and Social Council in its resolution 2000/35 of 18 October 2000, and recalling the decision of the Council, in its resolution 2006/49 of 28 July 2006, to strengthen the international arrangement on forests;

Reaffirming their commitment to the *Rio Declaration on Environment and Development*,[⑥] including that States have, in accordance with the *Charter of the United Nations* and the principles of international law, the sovereign right to exploit their own resources pursuant to their own environmental and developmental policies and the responsibility to ensure that activities within their jurisdiction or control do not cause damage to the environment of other States or of areas beyond the limits of national jurisdiction, and to the common but differentiated responsibilities of countries, as set out in Principle 7 of the *Rio Declaration*;

Recognizing that sustainable forest management, as a dynamic and evolving concept, is intended to maintain and enhance the economic, social and environmental value of all types of forests, for the benefit of present and future generations;

---

① Report of the United Nations Conference on Environment and Development, Rio de Janeiro, 3-14 June 1992, vol. I, Resolutions Adopted by the Conference (United Nations publication, Sales No. E.93.I.8 and corrigendum), resolution 1, annex III.

② Ibid., annex II.

③ Report of the World Summit on Sustainable Development, Johannesburg, South Africa, 26 August–4 September 2002 (United Nations publication, Sales No. E.03.II.A.1 and corrigendum), chap. I, resolution

④ Report of the International Conference on Financing for Development, Monterrey, Mexico, 18-22 March 2002 (United Nations publication, Sales No. E.02.II.A.7), chap. I, resolution 1, annex.

⑤ See General Assembly Resolution 60/1.

⑥ Report of the United Nations Conference on Environment and Development, Rio de Janeiro, 3-14 June 1992, vol. I, Resolutions Adopted by the Conference (United Nations publication, Sales No. E.93.I.8 and corrigendum), resolution 1, annex I.

Expressing their concern about continued deforestation and forest degradation, as well as the slow rate of afforestation and forest cover recovery and reforestation, and the resulting adverse impact on economies, the environment, including biological diversity, and the livelihoods of at least a billion people and their cultural heritage, and emphasizing the need for more effective implementation of sustainable forest management at all levels to address these critical challenges;

Recognizing the impact of climate change on forests and sustainable forest management, as well as the contribution of forests to addressing climate change;

Reaffirming the special needs and requirements of countries with fragile forest ecosystems, including those of low-forest-cover countries;

Stressing the need to strengthen political commitment and collective efforts at all levels, to include forests in national and international development agendas, to enhance national policy coordination and international cooperation and to promote intersectoral coordination at all levels for the effective implementation of sustainable management of all types of forests;

Emphasizing that effective implementation of sustainable forest management is critically dependent upon adequate resources, including financing, capacity development and the transfer of environmentally sound technologies, and recognizing in particular the need to mobilize increased financial resources, including from innovative sources, for developing countries, including least developed countries, landlocked developing countries and small island developing States, as well as countries with economies in transition;

Also emphasizing that implementation of sustainable forest management is also critically dependent upon good governance at all levels;

Noting that the provisions of this instrument do not prejudice the rights and obligations of Member States under international law;

Have committed themselves as follows:

I. Purpose

1. The purpose of this instrument is:

(a) To strengthen political commitment and action at all levels to implement effectively sustainable management of all types of forests and to achieve the shared global objectives on forests;

(b) To enhance the contribution of forests to the achievement of the internationally agreed development goals, including the *2030 Agenda for Sustainable Development*[①] and the *Sustainable Development Goals*[②];

(c) To provide a framework for national action and international cooperation.

II. Principles

2. Member States should respect the following principles, which build upon the *Rio*

---

① See General Assembly Resolution 70/1.

② See General Assembly Resolution 71/286, reference to Millennium Development Goals was revised to *2030 Agenda for Sustainable Development* and the *Sustainable Development Goals*.

Declaration on Environment and Development and the *Rio Forest Principles*:

(a) The instrument is voluntary and non-legally binding;

(b) Each State is responsible for the sustainable management of its forests and for the enforcement of its forest-related laws;

(c) Major groups as identified in *Agenda 21*,① local communities, forest owners and other relevant stakeholders contribute to achieving sustainable forest management and should be involved in a transparent and participatory way in forest decision-making processes that affect them, as well as in implementing sustainable forest management, in accordance with national legislation;

(d) Achieving sustainable forest management, in particular in developing countries as well as in countries with economies in transition, depends on significantly increased, new and additional financial resources from all sources;

(e) Achieving sustainable forest management also depends on good governance at all levels;

(f) International cooperation, including financial support, technology transfer, capacity-building and education, plays a crucial catalytic role in supporting the efforts of all countries, particularly developing countries as well as countries with economies in transition, to achieve sustainable forest management.

III. Scope

3. The present instrument applies to all types of forests.

4. Sustainable forest management, as a dynamic and evolving concept, aims to maintain and enhance the economic, social and environmental values of all types of forests, for the benefit of present and future generations.

IV. Global objectives on forests

5. Member States reaffirm the following shared global objectives on forests and their commitment to work globally, regionally and nationally to achieve progress towards their achievement by 2030②.

Global objective 1

Reverse the loss of forest cover worldwide through sustainable forest management, including protection, restoration, afforestation and reforestation, and increase efforts to prevent forest degradation;

Global objective 2

Enhance forest-based economic, social and environmental benefits, including by improving the livelihoods of forest-dependent people;

---

① The major groups identified in *Agenda 21* are women, children and youth, indigenous people and their communities, non-governmental organizations, local authorities, workers and trade unions, business and industry, scientific and technological communities, and farmers.

② See General Assembly Resolution 70/199, the timeline of the global objectives on forests is extended to 2030.

Global objective 3

Increase significantly the area of protected forests worldwide and other areas of sustainably managed forests, as well as the proportion of forest products from sustainably managed forests;

Global objective 4

Reverse the decline in official development assistance for sustainable forest management and mobilize significantly increased, new and additional financial resources from all sources for the implementation of sustainable forest management.

Ⅴ. **National policies and measures**

6. To achieve the purpose of the present instrument, and taking into account national policies, priorities, conditions and available resources, Member States should:

(a) Develop, implement, publish and, as necessary, update national forest programmes or other strategies for sustainable forest management which identify actions needed and contain measures, policies or specific goals, taking into account the relevant proposals for action of the Intergovernmental Panel on Forests/Intergovernmental Forum on Forests and resolutions of the United Nations Forum on Forests;

(b) Consider the seven thematic elements of sustainable forest management[①], which are drawn from the criteria identified by existing criteria and indicators processes, as a reference framework for sustainable forest management and, in this context, identify, as appropriate, specific environmental and other forest-related aspects within those elements for consideration as criteria and indicators for sustainable forest management;

(c) Promote the use of management tools to assess the impact on the environment of projects that may significantly affect forests, and promote good environmental practices for such projects;

(d) Develop and implement policies that encourage the sustainable management of forests to provide a wide range of goods and services and that also contribute to poverty reduction and the development of rural communities;

(e) Promote efficient production and processing of forest products, with a view, inter alia, to reducing waste and enhancing recycling;

(f) Support the protection and use of traditional forest-related knowledge and practices in sustainable forest management with the approval and involvement of the holders of such knowledge, and promote fair and equitable sharing of benefits from their utilization, in accordance with national legislation and relevant international agreements;

(g) Further develop and implement criteria and indicators for sustainable forest management that are consistent with national priorities and conditions;

(h) Create enabling environments to encourage private-sector investment, as well as

---

① The elements are (ⅰ) extent of forest resources; (ⅱ) forest biological diversity; (ⅲ) forest health and vitality; (ⅳ) productive functions of forest resources; (ⅴ) protective functions of forest resources; (ⅵ) socio-economic functions of forests; and (ⅶ) legal, policy and institutional framework.

investment by and involvement of local and indigenous communities, other forest users and forest owners and other relevant stakeholders, in sustainable forest management, through a framework of policies, incentives and regulations;

(i) Develop financing strategies that outline the short-, medium- and long-term financial planning for achieving sustainable forest management, taking into account domestic, private-sector and foreign funding sources;

(j) Encourage recognition of the range of values derived from goods and services provided by all types of forests and trees outside forests, as well as ways to reflect such values in the marketplace, consistent with relevant national legislation and policies;

(k) Identify and implement measures to enhance cooperation and cross-sectoral policy and programme coordination among sectors affecting and affected by forest policies and management, with a view to integrating the forest sector into national decision-making processes and promoting sustainable forest management, including by addressing the underlying causes of deforestation and forest degradation, and by promoting forest conservation;

(l) Integrate national forest programmes, or other strategies for sustainable forest management, as referred to in paragraph 6(a) above, into national strategies for sustainable development, relevant national action plans and poverty-reduction strategies;

(m) Establish or strengthen partnerships, including public-private partnerships, and joint programmes with stakeholders to advance the implementation of sustainable forest management;

(n) Review and, as needed, improve forest-related legislation, strengthen forest law enforcement and promote good governance at all levels in order to support sustainable forest management, to create an enabling environment for forest investment and to combat and eradicate illegal practices, in accordance with national legislation, in the forest and other related sectors;

(o) Analyse the causes of, and address solutions to, threats to forest health and vitality from natural disasters and human activities, including threats from fire, pollution, pests, disease and invasive alien species;

(p) Create, develop or expand, and maintain networks of protected forest areas, taking into account the importance of conserving representative forests, by means of a range of conservation mechanisms, applied within and outside protected forest areas;

(q) Assess the conditions and management effectiveness of existing protected forest areas with a view to identifying improvements needed;

(r) Strengthen the contribution of science and research in advancing sustainable forest management by incorporating scientific expertise into forest policies and programmes;

(s) Promote the development and application of scientific and technological innovations, including those that can be used by forest owners and local and indigenous communities to advance sustainable forest management;

(t) Promote and strengthen public understanding of the importance of and the benefits provided by forests and sustainable forest management, including through public awareness programmes and education;

(u) Promote and encourage access to formal and informal education, extension and training programmes on the implementation of sustainable forest management;

(v) Support education, training and extension programmes involving local and indigenous communities, forest workers and forest owners, in order to develop resource management approaches that will reduce the pressure on forests, particularly fragile ecosystems;

(w) Promote active and effective participation by major groups, local communities, forest owners and other relevant stakeholders in the development, implementation and assessment of forest-related national policies, measures and programmes;

(x) Encourage the private sector, civil society organizations and forest owners to develop, promote and implement in a transparent manner voluntary instruments, such as voluntary certification systems or other appropriate mechanisms, to develop and promote forest products from sustainably managed forests harvested in accordance with domestic legislation, and to improve market transparency;

(y) Enhance access by households, small-scale forest owners, forest-dependent local and indigenous communities, living in and outside forest areas, to forest resources and relevant markets in order to support livelihoods and income diversification from forest management, consistent with sustainable forest management.

VI. International cooperation and means of implementation

7. To achieve the purpose of the present instrument, Member States should:

(a) Make concerted efforts to secure a sustained high-level political commitment to strengthen the means of implementation of sustainable forest management, including financial resources, to provide support, in particular for developing countries and countries with economies in transition, as well as to mobilize and provide significantly increased, new and additional financial resources from private, public, domestic and international sources to and within developing countries, as well as countries with economies in transition;

(b) Reverse the decline in official development assistance for sustainable forest management and mobilize significantly increased, new and additional financial resources from all sources for the implementation of sustainable forest management;

(c) Take action to raise the priority of sustainable forest management in national development plans and other plans, including poverty-reduction strategies, in order to facilitate increased allocation of official development assistance and financial resources from other sources for sustainable forest management;

(d) Develop and establish positive incentives, in particular for developing countries as well as countries with economies in transition, to reduce the loss of forests, to promote reforestation, afforestation and rehabilitation of degraded forests, to implement sustainable forest management and to increase the area of protected forests;

(e) Support the efforts of countries, particularly developing countries as well as countries with economies in transition, to develop and implement economically, socially and environmentally sound measures that act as incentives for the sustainable management of forests;

(f) Strengthen the capacity of countries, in particular developing countries, to significantly increase the production of forest products from sustainably managed forests;

(g) Enhance bilateral, regional and international cooperation with a view to promoting international trade in forest products from sustainably managed forests harvested according to domestic legislation;

(h) Enhance bilateral, regional and international cooperation to address illicit international trafficking in forest products through the promotion of forest law enforcement and good governance at all levels;

(i) Strengthen, through enhanced bilateral, regional and international cooperation, the capacity of countries to combat effectively illicit international trafficking in forest products, including timber, wildlife and other forest biological resources;

(j) Strengthen the capacity of countries to address forest-related illegal practices, including wildlife poaching, in accordance with domestic legislation, through enhanced public awareness, education, institutional capacity-building, technological transfer and technical cooperation, law enforcement and information networks;

(k) Enhance and facilitate access to and transfer of appropriate, environmentally sound and innovative technologies and corresponding know-how relevant to sustainable forest management and to efficient value-added processing of forest products, in particular to developing countries, for the benefit of local and indigenous communities;

(l) Strengthen mechanisms that enhance sharing among countries and the use of best practices in sustainable forest management, including through freeware-based information and communications technology;

(m) Strengthen national and local capacity in keeping with their conditions for the development and adaptation of forest-related technologies, including technologies for the use of fuelwood;

(n) Promote international technical and scientific cooperation, including South-South cooperation and triangular cooperation, in the field of sustainable forest management, through the appropriate international, regional and national institutions and processes;

(o) Enhance the research and scientific forest-related capacities of developing countries and countries with economies in transition, particularly the capacity of research organizations to generate and have access to forest-related data and information, and promote and support integrated and interdisciplinary research on forest-related issues, and disseminate research results;

(p) Strengthen forestry research and development in all regions, particularly in developing countries and countries with economies in transition, through relevant organizations, institutions and centres of excellence, as well as through global, regional and subregional networks;

(q) Strengthen cooperation and partnerships at the regional and subregional levels to promote sustainable forest management;

(r) As members of the governing bodies of the organizations that form the Collaborative Partnership on Forests, help ensure that the forest-related priorities and programmes of members of the Partnership are integrated and mutually supportive, consistent with their mandates, taking into account relevant policy recommendations of the United Nations Forum on Forests;

(s) Support the efforts of the Collaborative Partnership on Forests to develop and implement joint initiatives.

Ⅶ. **Monitoring, assessment and reporting**

8. Member States should monitor and assess progress towards achieving the purpose of the present instrument.

9. Member States should submit, on a voluntary basis, taking into account the availability of resources and the requirements and conditions for the preparation of reports for other bodies or instruments, national progress reports as part of their regular reporting to the Forum.

Ⅷ. **Working modalities**

10. The Forum should address, within the context of its multi-year programme of work, the implementation of the present instrument.

Notes

1. The *Non-legally Binding Authoritative Statement of Principles for a Global Consensus on Management, Conservation and Sustainable Development of All Types of Forests (Forest Principles)*

《关于所有类型森林的管理、保存和可持续开发的无法律约束力的全球协商一致意见权威性原则声明》，简称《有关森林问题的原则声明》。于 1992 年 6 月 3 日至 14 日，在巴西里约热内卢召开的联合国环境与发展会议通过。

2. Intergovernmental Panel on Forests/Intergovernmental Forum on Forests; the United Nations Forum on Forests

政府间森林问题特设小组(Intergovernmental Panel on Forests，IPF) 成立于 1995 年，是 1995—1997 年政府间森林论坛（Intergovernmental Forum on Forests，IFF）的前身机构。为了执行联合国环境与发展会议所作决定的后续行动，根据 1995 年 6 月 1 日联合国经联合国经济及社会理事会（Economic and Social Council）第 1995/226 号决定所设立，重点是执行联合国环境与发展会议(环发会议)与森林有关的决定，目的是继续政府间森林政策对话，以寻求协商一致意见和协调的行动建议，打击滥伐森林、遏制森林退化并支持所有类型的森林的管理、养护和可持续发展。

政府间森林论坛（Intergovernmental Forum on Forests，IFF）建立于 1997 年联合国大会，在联合国可持续发展委员会(CSD)下运作，继续就森林问题进行政府间政策对话，并致力于就森林问题的国际安排和机制达成协议，包括一项具有法律约束力的条约。IFF 在 1997 年 10 月至 2000 年 2 月举行了 4 次会议，共提出约 120 项行动建议。

在 2000 年 2 月召开的第 4 届，也是最后一届会议期间提出了建立联合国森林论坛，邀请相关国际组织、机构和联合国组织参与森林合作伙伴关系的建议。2000 年 10 月 18 日，联合国经济及社会理事会(经社理事会)设立了联合国森林论坛（United Nations Forum on Forests）——一个拥有全球会员的、高级别的政府间组织。其主要目标是促进对各类森林进行的管理、保护和可持续发展，加强为此目的做出的政治承诺。

3. The *Johannesburg Declaration on Sustainable Development* and the *Plan of Implementation of the World Summit on Sustainable Development*

《约翰内斯堡可持续发展承诺》和《执行计划》，是 2002 年 9 月在南非首都约翰内斯堡举行的可持续发展世界首脑会议上通过的两个文件。这次会议是 1992 年里约地球峰会的后续，会议召开之时，正值世界范围内贫富分化日趋严重，人类在健康、生物多样性、农业生产、水和能源五大领域面临严峻挑战，全球可持续发展状况有恶化的趋势。因而，在作为会议政治宣言的《约翰内斯堡可持续发展承诺》中，各国承诺将不遗余力地执行可持续发展战略，把世界建成一个以人为本、人类与自然协调发展的美好社会。《执行计划》则指出，当今世界面临的最严峻的全球性挑战是贫困，消除贫困是全球可持续发展必不可少的条件。把消除贫困纳入可持续发展理念之中，并作为这次峰会的主旋律之一，标志着人类可持续发展理念提高到了一个新的层次。

4. The *Monterrey Consensus of the International Conference on Financing for Development*

《发展筹资问题国际会议蒙特雷共识》，简称《蒙特雷共识》。2002 年 3 月，为期 5 天的联合国发展筹资国际会议 22 日在墨西哥北部工业城市蒙特雷落下帷幕。各国家元首或政府首脑就国际发展筹资达成共识，即《蒙特雷共识》。《蒙特雷共识》主要包括调动国内经济资源、增加私人国际投资、开放市场和确保公平的贸易体制、增加官方发展援助、解决发展中国家的债务困难和改善全球和区域金融结构、发展中国家在国际决策中的公正代表性 6 个方面内容。《蒙特雷共识》指出：发达国家和发展中国家应该建立一种新的伙伴关系，全面落实《联合国千年宣言》中提出的旨在实现消除贫困、改善社会状况、提高生活水平和保护环境等各项可持续发展目标。

Key Words and Phrases

1. afforestation /ˌæfɒrɪˈsteɪʃn/ n. (technical) the process of planting areas of land with trees in order to form a forest 造林

2. allocation /ˌæləˈkeɪʃ(ə)n/ n. the act of distributing by allotting or apportioning or according to a plan 分配；配置；安置

3. binding /ˈbaɪndɪŋ/ adj. ~ (on/upon sb.) that must be obeyed because it is accepted in law 装订；捆绑

4. catalytic /ˌkætəˈlɪtɪk/ adj. cause things to happen or they increase

the speed at which things happen 接触反应的；起催化作用的

5. consensus	/kən'sensəs/	n.	an opinion that all members of a group agree with 共识；一致；同意
6. degradation	/ˌdegrə'deɪʃn/	n.	(technical) the process of sth. being damaged or made worse 退化；降级
7. evolve	/ɪ'vɒlv/	vt. & vi.	gradually develops over a period of time into something different and usually more advanced 发展；进化；使逐步形成；推断出
8. expertise	/ˌekspɜː'tiːz/	n.	expert knowledge or skill in a particular subject, activity or job 专长；专门知识；专门技术；专家的意见
9. exploit	/ɪk'splɔɪt/	vt.	to use sth. well in order to gain as much from it as possible 开发；开采
10. forum	/'fɔːrəm/	n.	a place where people can exchange opinions and ideas on a particular issue or a meeting organized for this purpose 论坛；讨论会
11. fragile	/'frædʒaɪl/	adj.	easily broken or damaged; weak and uncertain; easily destroyed or spoilt 脆的；易碎的
12. illicit	/ɪ'lɪsɪt/	adj.	contrary to or forbidden by law or the social customs of a country 违法的；不正当的
13. inception	/ɪn'sepʃ(ə)n/	n.	(sing.) (formal) the start of an institution, an organization, etc. （机构、组织等的）开端；创始
14. incentive	/ɪn'sentɪv/	n.	something that encourages you to do sth. 激励；刺激；鼓励
15. instrument	/'ɪnstrəmənt/	n.	(law) a document that states some contractual relationship or grants some right 仪器；器具；正式法律文件
16. livelihood	/'laɪvlihʊd/	n.	(usually sing. U) a means of earning money in order to live 生计
17. mobilize	/'məʊbəlaɪz/	vt. & vi.	encouraging people to take action, especially political action 鼓动；动员；

				调动
18.	reforestation	/ˌriːfɒrɪˈsteɪʃn/	n.	the restoration (replanting) of a forest that had been reduced by fire or cutting 重新造林；再造林
19.	resolution	/ˌrezəˈluːʃ(ə)n/	n.	a formal statement of an opinion agreed on by a committee or a council, especially by means of a vote 决议；正式决定；解决
20.	reverse	/rɪˈvɜːs/	vt. & vi.	to change sth. completely so that it is the opposite of what it was before 颠倒；彻底转变；撤销，废除（决定、法律等）
21.	stakeholder	/ˈsteɪkhəʊldə(r)/	n.	a person who has an interest in a company's or organization's affairs 利益相关者；（某组织、工程、体系等的）参与人
22.	trafficking	/ˈtræfɪkɪŋ/	n.	buying and selling sth. illegally 非法交易（尤指毒品买卖）
23.	transparent	/trænsˈpærənt/	adj.	easily understood or recognized 透明的；清澈的；显而易见的
24.	utilization	/ˌjuːtəlaɪˈzeɪʃn/	n.	the act of using, the state of having been made use of 利用；使用
25.	deforestation	/ˌdiːˌfɒrɪˈsteɪʃn/	n.	the clearing or severe thinning of a forest or other wooded area, leaving few or no trees 毁林；滥伐森林
26.	non-legally binding			无法律约束力
27.	sustainable forest management			森林可持续经营
28.	forest degradation			森林退化

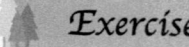

Exercises

Exercise 1   Reading Comprehension

Directions: Read *the Non-legally Binding Instrument on All Types of Forests*, and decide whether the following statements are true or false. Write T for true or F for false in the brackets in front of each statement.

1. (   ) As a dynamic and evolving concept, sustainable forest management aims to maintain and enhance the economic, social and environmental values of all types of forests, for the benefit of present and future generations.

2. (　　) Thanks to the international concern over deforestation and forest degradation, the world is witnessing a higher rate of afforestation, forest cover recovery and reforestation.

3. (　　) It is vital for developed countries to mobilize increased financial resources to enable effective implementation of sustainable forest management.

4. (　　) Major groups as identified in *the Forest Instrument*, local communities, forest owners and other relevant stakeholders should be involved in forest decision-making processes and in implementing sustainable forest management in accordance with national legislation.

5. (　　) Traditional forest-related knowledge and practices should be replaced by modern approaches to sustainable forest management.

6. (　　) Bilateral, regional and international cooperation must be enhanced to address illicit international trafficking in forest products through the promotion of forest law enforcement and good governance at all levels.

Exercise 2　Skimming and Scanning

Directions: *Read the following passage excerpted from* **United Nations Forest Instrument**. *At the end of the passage, there are six statements. Each statement contains information given in one of the paragraphs of the passage. Identify the paragraph from which the information is derived. Each paragraph is marked with a letter. You may choose a paragraph more than once. Answer the questions by writing the corresponding letter in the brackets in front of each statement.*

A) To achieve the purpose of the present instrument, and taking into account national policies, priorities, conditions and available resources, Member States should;

B) Develop, implement, publish and, as necessary, update national forest programmes or other strategies for sustainable forest management which identify actions needed and contain measures, policies or specific goals, taking into account the relevant proposals for action of the Intergovernmental Panel on Forests/Intergovernmental Forum on Forests and resolutions of the United Nations Forum on Forests;

C) Consider the seven thematic elements of sustainable forest management, which are drawn from the criteria identified by existing criteria and indicators processes, as a reference framework for sustainable forest management and, in this context, identify, as appropriate, specific environmental and other forest-related aspects within those elements for consideration as criteria and indicators for sustainable forest management;

D) Support the protection and use of traditional forest-related knowledge and practices in sustainable forest management with the approval and involvement of the holders of such knowledge, and promote fair and equitable sharing of benefits from their utilization, in accordance with national legislation and relevant international agreements;

E) Identify and implement measures to enhance cooperation and cross-sectoral policy and programme coordination among sectors affecting and affected by forest policies and management, with a view to integrating the forest sector into national decision-making processes and promoting sustainable forest management, including by addressing the underlying causes of deforestation and forest degradation, and by promoting forest conservation;

F) Review and, as needed, improve forest-related legislation, strengthen forest law enforcement and promote good governance at all levels in order to support sustainable forest management, to create an enabling environment for forest investment and to combat and eradicate illegal practices, in accordance with national legislation, in the forest and other related sectors;

G) Create, develop or expand, and maintain networks of protected forest areas, taking into account the importance of conserving representative forests, by means of a range of conservation mechanisms, applied within and outside protected forest areas;

H) Support education, training and extension programmes involving local and indigenous communities, forest workers and forest owners, in order to develop resource management approaches that will reduce the pressure on forests, particularly fragile ecosystems;

I) Encourage the private sector, civil society organizations and forest owners to develop, promote and implement in a transparent manner voluntary instruments, such as voluntary certification systems or other appropriate mechanisms, to develop and promote forest products from sustainably managed forests harvested in accordance with domestic legislation, and to improve market transparency;

J) Enhance access by households, small-scale forest owners, forest-dependent local and indigenous communities, living in and outside forest areas, to forest resources and relevant markets in order to support livelihoods and income diversification from forest management, consistent with sustainable forest management.

1. (   ) Traditional forest-related knowledge and practices in sustainable forest management will be protected and used with the approval and involvement of the holders of such knowledge.

2. (   ) An enabling environment shall be created for forest investment to combat and eradicate illegal practices in the forest and other related sectors.

3. (   ) National forest programmes shall be developed, implemented, published and necessarily updated to achieve the purpose of the present instrument.

4. (   ) To develop and promote forest products, voluntary certification systems or other appropriate mechanisms shall be developed, promoted and implemented in a transparent manner.

5. (   ) The forest sector shall be integrated into national decision-making processes to promote sustainable forest management by addressing the underlying causes of deforestation and forest degradation.

6. (   ) From the criteria identified by existing criteria and indicators processes, seven thematic elements of sustainable forest management are drawn as a reference framework for sustainable forest management.

Exercise 3 Word Formation

Directions: *In this section, there are ten sentences from **United Nations Forest Instrument**. You are required to complete these sentences with the proper form of the words given in blanks.*

1. Major groups should be involved in a transparent and _____ way in forest decision-making processes that affect them. (participate)

2. To achieve the purpose of the present instrument, Member States should consider the seven _____ elements of sustainable forest management, which are drawn from the criteria identified by existing criteria and indicators processes. (theme)

3. Member states should further develop and implement criteria and _____ for sustainable forest management that are consistent with national priorities and conditions. (indicate)

4. Member states should encourage _____ of the range of values derived from goods and services provided by all types of forests and trees outside forests. (recognize)

5. Review and, as needed, improve forest-related legislation, strengthen forest law enforcement and promote good _____ at all levels in order to support sustainable forest management. (govern)

6. Enhance access by households, small-scale forest owners, forest-dependent local and indigenous communities, living in and outside forest areas, to forest resources and relevant markets in order to support livelihoods and income _____ from forest management. (diverse)

7. Reverse the decline in official development assistance for sustainable forest management and _____ significantly increased, new and additional financial resources from all sources for the implementation of sustainable forest management. (mobile)

8. Strengthen the capacity of countries to combat effectively illicit international trafficking in forest products, including timber, wildlife and other forest _____ resources. (biology)

9. Promote international technical and scientific cooperation, including South-South cooperation and _____ cooperation, in the field of sustainable forest management. (triangle)

10. As members of the governing bodies of the organizations that form the Collaborative Partnership on Forests, help ensure that the forest-related _____ and programmes of members of the Partnership are integrated and mutually supportive. (prior)

Exercise 4 Translation

Section A

Directions: *Read **United Nations Forest Instrument**, and complete the sentences by translating into English the Chinese given in blanks.*

1. Recognizing that _____ (可持续森林经营), as a

dynamic and evolving concept, is intended to maintain and enhance the economic, social and environmental value of all types of forests, for the benefit of present and future generations.

2. Reverse the loss of forest cover worldwide through sustainable forest management, including protection, restoration, afforestation and reforestation, and increase efforts to prevent _____ (森林退化).

3. International cooperation, including financial support, technology transfer, capacity-building and education, plays a crucial catalytic role in supporting the efforts of all countries, particularly developing countries as well as _____ (经济转型国家), to achieve sustainable forest management.

4. Create enabling environments to encourage private-sector investment, as well as investment by and involvement of _____ (地方和土著社区), other forest users and forest owners and other relevant stakeholders, in sustainable forest management, through a framework of policies, incentives and regulations.

5. Identify and implement measures to enhance cooperation and cross-sectoral policy and programme coordination among sectors affecting and affected by forest policies and management, with a view to integrating the forest sector into national decision-making processes and promoting sustainable forest management, including by addressing the underlying causes of _____ (滥伐森林和森林退化), and by promoting forest conservation.

6. Review and, as needed, improve forest-related legislation, strengthen forest law enforcement and promote good governance at all levels in order to support sustainable forest management, create_____ (一个有利于森林投资的环境) and to combat and eradicate illegal practices, in accordance with national legislation, in the forest and other related sectors.

7. Support education, training and extension programmes involving local and indigenous communities, forest workers and forest owners, in order to develop resource management approaches that will reduce _____ (森林，尤其是脆弱生态系统承受的压力).

8. Develop and establish positive incentives, in particular for developing countries as well as countries with economies in transition, to reduce the loss of forests, to promote _____ (重新造林、植树造林和已退化的森林的恢复)，to implement sustainable forest management and to increase the area of protected forests.

9. Strengthen, through enhanced bilateral, regional and international cooperation, the capacity of countries to combat effectively _____ (非法国际贩运行为) in forest products, including timber, wildlife and other forest biological resources.

10. As members of the governing bodies of the organizations that form _____ (森林合作伙伴关系), help ensure that the forest-related priorities and programmes of members of the Partnership are integrated and mutually supportive, consistent

with their mandates, taking into account relevant policy recommendations of the United Nations Forum on Forests.

Section B

Directions: *Translate the following sentences from English into Chinese.*

1. Reaffirming their commitment to the *Rio Declaration on Environment and Development*, including that States have, in accordance with the *Charter of the United Nations* and the principles of international law, the sovereign right to exploit their own resources pursuant to their own environmental and developmental policies and the responsibility to ensure that activities within their jurisdiction or control do not cause damage to the environment of other States or of areas beyond the limits of national jurisdiction, and to the common but differentiated responsibilities of countries, as set out in *Principle of the Rio Declaration*. (*Preamble*)

2. Major groups as identified in *Agenda 21*, local communities, forest owners and other relevant stakeholders contribute to achieving sustainable forest management and should be involved in a transparent and participatory way in forest decision-making processes that affect them, as well as in implementing sustainable forest management, in accordance with national legislation. (*Principles of the NLBI*)

3. Achieving sustainable forest management, in particular in developing countries as well as in countries with economies in transition, depends on significantly increased, new and additional financial resources from all sources. (*Principles of the NLBI*)

4. Develop, implement, publish and, as necessary, update national forest programmes or other strategies for sustainable forest management which identify actions needed and contain measures, policies or specific goals, taking into account the relevant proposals for action of the Intergovernmental Panel on Forests/Intergovernmental Forum on Forests and resolutions of the United Nations Forum on Forests. (*National policies and measures*)

5. Identify and implement measures to enhance cooperation and cross-sectoral policy and programme coordination among sectors affecting and affected by forest policies and management, with a view to integrating the forest sector into national decision-making processes and promoting sustainable forest management, including by addressing the underlying causes of deforestation and forest degradation, and by promoting forest conservation. (*National policies and measures*)

6. Make concerted efforts to secure a sustained high-level political commitment to strengthen the means of implementation of sustainable forest management, including financial resources, to provide support, in particular for developing countries and countries with economies in transition, as well as to mobilize and provide significantly increased, new and additional financial resources from private, public, domestic and international sources to and within developing countries, as well as countries with economies in transition. (*International cooperation and means of implementation*)

Passage 1

Directions: *Read the following passage and choose the best answer for each of the following questions according to the information given in the passage.*

Most countries are already implementing forest development plans and programmes at various scales that include at least some of the policy measures adopted in *the Forest Instrument*. For example, many developing countries are implementing community-based forest management programmes as part of their national forest programmes (nfps) whilst others are part of the FLEGT (Forest Law Enforcement, Governance and Trade) process. These programmes directly contribute to the implementation of the policy measures on:"enhancing access of households, small scale forest owners and forest dependent communities to forest resources and improving forest legislation and strengthening forest law enforcement" respectively.

The national policies and measures represent a broad and comprehensive set of measures required at national level to achieve sustainable forest management. However, the majority of the countries have not yet conducted a deliberate assessment of where they are in relation to implementation of *the Forest Instrument*'s national policies and measures or assessed progress through monitoring. In order to facilitate implementation of the NLBI, it would be useful if countries could systematically relate their National Forest Programme (nfp) to *the Forest Instrument*, and identify those areas of *the Forest Instrument* that are inadequately addressed by their current national policies, strategies and ongoing forestry-related development initiatives. This would provide them with a basis for prioritizing areas for incorporating into their national forest programmes in order to advance their efforts towards achieving sustainable forest management. At the country level most forestry stakeholders are not familiar with *the Forest Instrument*, its purpose and objectives. Stakeholders in the countries need to gain a better understanding of the forest instrument and of its usefulness in achieving progress in their efforts to sustainably manage their forest resources. In particular implementing the NLBI provides a country with an opportunity to view and implement forestry initiatives in a comprehensive and systematic manner.

In this regard the expected benefits include:
• Heightened political commitment to sustainable forest management (SFM) at national level.
• An overarching framework for forestry development.
• Improved environmental, social and economic contribution of forests at national level and to internationally agreed development and environmental sustainability goals.
• A single framework for coordination of national and international forestry actions.
• A holistic and comprehensive "360 degree" view of forests that reduces fragmentation

of efforts and programmes.
  • A tool for assessing progress towards sustainable forest management at national level.
Some of the benefits are described in detail below:
  a) Taking the NLBI as an overarching forestry development framework

The NLBI can serve as an overarching framework for facilitating integration of national and international forestry-related policies and programmes at national level. It enhances coordination among various forest-related policy processes and programmes which, in most countries, are often implemented in a fragmented manner. The fragmentation often leads to duplication and inefficient use scarce resources. In the worst case it can even lead to contradictory policies, legislation and programmes. Coordination is particularly important for international forest and related initiatives such as forest biodiversity conservation, reducing emissions from deforestation and forest degradation (REDD+), Forest Law Enforcement, Governance and Trade (FLEGT) and related to this, the Voluntary Partnership Agreements (VPA) between the EU, and partner countries and combating desertification. Coordination is also important with other forest-relevant sectors of the national economy.

  b) Linking the NLBI to national development frameworks

Implementation of the NLBI provides an opportunity of integrating forest related policies and programmes into national development frameworks such as poverty reduction strategies, NAPAS and NEMAs. This can be achieved by incorporating the priority policies and measures into the national forest programmes and ensuring these are properly integrated into the national development programmes. There are several approaches that can be used to ensure integration into national development plans. The first approach is to clearly articulate the contribution of forests to the national economy and then ensuring this is reflected in the periodic national development plans. For example, Uganda has identified forestry as one of the four drivers of economic growth in its current five-year national development plan. In addition, they recently took advantage of the drafting of a new constitution to incorporate forestry into their national development plan with a target of reaching 10% forest cover in the next 20 years.

  c) Linking with and coordinating international forestry related agreements and development goals

There are many international forestry-related processes that are being implemented by national governments especially under the three environmental conventions namely CBD, UNFCCC and UNCCD. Countries can use implementation of the NLBI to facilitate and coordinate implementation of forestry activities under the international conventions.

  d) Taking the NLBI as a framework for enhancing inter-sectoral coordination

There are many forest-related processes and initiatives that are being implemented in other sectors such as energy, water, agriculture and environment. By conducting an inventory of all on-going forestry related actives, using the FI national measures as a checklist, and identifying the stakeholders involved, implementation of the FI can provide a basis for effective cross-sectoral coordination. This is essential for harnessing synergies and avoiding duplication and sometimes contradictory policies and initiatives at country level.

e) A basis for resource mobilisation

By deliberately integrating the elements of the NLBI into the nfp and clearly demonstrating the link between the nfp and the NLBI, and in particular the contribution to the global objectives of achieving SFM and the global objectives on forests, countries can have a basis for developing a comprehensive financing strategy. Experience to-date has shown that most developing countries have difficulties implementing their nfps due to lack of financial resources and institutional capacity. To address these challenges the countries are developing comprehensive financing strategies to mobilise financial and technical resources from all sources especially domestic public financing, private sector investment and bilateral and multilateral support. Where the contribution of the nfp to the NLBI and the global objectives on forests is clearly articulated it may be easier to guide and convince development partners to contribute to the funding of the nfp.

1. What is the main achievement for most countries in implementing *the Forest Instrument*? _____

   A. They could systematically relate their National Forest Programme to *the Forest Instrument*.
   B. They have identified those areas of *the Forest Instrument* that are inadequately addressed by their current national policies, strategies and ongoing forestry-related development initiatives.
   C. They have conducted a deliberate assessment of where they are in relation to implementation of *the Forest Instrument*'s national policies.
   D. They have carried out forest development plans and programmes at various scales in *the Forest Instrument*.

2. What is stressed by the author about most forestry stakeholders? _____

   A. Stakeholders lack knowledge about *the Forest Instrument*, its purpose and objectives.
   B. Stakeholders should be prioritized in order to achieve sustainable forest management.
   C. Most stakeholders in the countries gain a better understanding of *the Forest Instrument* and of its usefulness in achieving progress in their efforts to sustainably manage their forest resources.
   D. The NLBI provides most stakeholders with an opportunity to view and implement forestry initiatives in a comprehensive and systematic manner.

3. How many environmental conventions are mentioned in addressing employing the implementation of the NLBI to facilitate and coordinate implementation of forestry activities? _____

   A. 2   B. 3   C. 4   D. 5

4. Which of the followings is **NOT** the approach that can be used to ensure integration into national development plans? _____

   A. To clearly articulate the contribution of forests to the national economy.
   B. To ensure the contribution of forests reflected in the periodic national development plans.
   C. To enhance coordination among various forest-related policy processes and

programmes in a fragmented manner.

D. To take advantage of the drafting of a new constitution to incorporate forestry into their national development plan.

5. Experience to-date has shown that most developing countries have difficulties implementing their nfps due to _____.

A. lack of national development frameworks
B. lack of international forestry related agreements
C. lack of the support from forestry stakeholders
D. lack of financial resources and institutional capacity

Passage 2

Directions: *In this section, there is a passage with twelve blanks. You are required to select one word for each blank from a list of choices given in a word bank following the passage. Read the passage through carefully before making your choices. Each choice in the bank is identified by a letter. You may not use any of the words in the bank more than once.*

A. contribute	B. degradation	C. dynamic	D. ecosystem
E. environmental	F. implementation	G. legislation	H. ranging
I. species	J. sustainably	K. stakeholders	L. thematic

In its broadest sense, sustainable forest management (SFM) encompasses the administrative, legal, technical, economic, social and environmental aspects of the conservation and use of forests. It implies various degrees of human intervention, __1__ from actions aimed at safeguarding and maintaining forest ecosystems and their functions to those favouring specific socially or economically valuable __2__ or groups of species for the improved production of goods and services. In addition to forest products (comprising both wood and non-wood forest products), __3__ managed forests provide important ecosystem services, such as carbon sequestration, biodiversity conservation, and the protection of water resources.

Many of the world's forests and woodlands are not being managed sustainably, especially in the tropics and subtropics. Many countries lack appropriate forest __4__ regulation and incentives to promote SFM. Many have inadequate funding and human resources for the preparation, __5__ and monitoring of forest management plans and lack mechanisms to ensure the participation and involvement of all __6__ in forest governance, planning and development. Where forest management plans exist, they are frequently limited to ensuring the sustained production of wood and lack sufficient attention to the sustainable production of non-wood products and __7__ services and the maintenance of social and environmental values. Also, other land uses may appear more economically attractive to land managers (at least in the short term) than forest management, thus leading to forest __8__ and deforestation.

The aim of sustainable forest management is to ensure that forests supply goods and services to meet both present-day and future needs and __9__ to the sustainable development of communities. The United Nations General Assembly in 2007 recognizes SFM as a __10__ and evolving concept that aims to maintain and enhance the economic, social and __11__ values of all types of forests for the benefit of present and future generations, considering the following seven __12__ elements as a reference framework: ①extent of forest resources; ②forest biodiversity; ③forest health and vitality; ④productive functions of forest resources; ⑤protective functions of forest resources; ⑥socio-economic functions of forests; ⑦legal, policy and institutional framework.

Further Studies and Post-Reading Discussion

Task 1

Directions: *Surf the Internet and find more information about* **United Nations Forest Instrument**. *Work in groups and work out a report on one of the following topics.*

1. Purposes of *United Nations Forest Instrument*.
2. Global objectives on forests.
3. China's efforts in sustainable forest management.

Task 2

Directions: *Read the following sentences on Eco-Civilization and make a speech on your understanding of the eco-environmental conservation.*

人类与自然命运共同体

人类进入工业文明时代以来，在创造巨大物质财富的同时，也加速了对自然资源的攫取，打破了地球生态系统平衡，人与自然深层次矛盾日益显现。近年来，气候变化、生物多样性丧失、荒漠化加剧、极端气候事件频发，给人类生存和发展带来严峻挑战。新冠肺炎疫情持续蔓延，使各国经济社会发展雪上加霜。面对全球环境治理前所未有的困难，国际社会要以前所未有的雄心和行动，勇于担当，勠力同心，共同构建人与自然生命共同体。

坚持人与自然和谐共生。"万物各得其和以生，各得其

Since the time of the industrial civilization, mankind has created massive material wealth. Yet, it has come at a cost of intensified exploitation of natural resources, which disrupted the balance in the Earth's ecosystem, and laid bare the growing tensions in the human-Nature relationship. In recent years, climate change, biodiversity loss, worsening desertification and frequent extreme weather events have all posed severe challenges to human survival and development. The ongoing COVID-19 pandemic has added difficulty to economic and social development across countries. Faced with unprecedented challenges in global environmental governance, the international community needs to come up with unprecedented ambition and action. We need to act with a sense of responsibility and unity, and work together to foster a community of life for man and Nature.

We must be committed to harmony between man and Nature. "All things that grow live in harmony and

养以成。"大自然是包括人在内一切生物的摇篮，是人类赖以生存发展的基本条件。大自然孕育抚养了人类，人类应该以自然为根，尊重自然、顺应自然、保护自然。不尊重自然，违背自然规律，只会遭到自然报复。自然遭到系统性破坏，人类生存发展就成了无源之水、无本之木。我们要像保护眼睛一样保护自然和生态环境，推动形成人与自然和谐共生新格局。（摘自 2021 年 4 月《习近平在"领导人气候峰会"上的讲话》）	benefit from the nourishment of Nature." Mother Nature is the cradle of all living beings, including humans. It provides everything essential for humanity to survive and thrive. Mother Nature has nourished us, and we must treat Nature as our root, respect it, protect it, and follow its laws. Failure to respect Nature or follow its laws will only invite its revenge. Systemic spoil of Nature will take away the foundation of human survival and development, and will leave us human beings like a river without a source and a tree without its roots. We should protect Nature and preserve the environment like we protect our eyes, and endeavor to foster a new relationship where man and Nature can both prosper and live in harmony. (Excerpt from Remarks by H. E. Xi Jinping: *For Man and Nature: Building a Community of Life Together*)

第5章 巴黎协定

Chapter 5　Paris Agreement

Background and Significance

《巴黎协定》是由全世界178个缔约方共同签署的气候变化协定,为2020年后全球应对气候变化的行动作出统一安排。《巴黎协定》于2015年12月12日在第21届联合国气候变化大会(巴黎气候大会)上通过,2016年4月22日在美国纽约联合国大厦签署,2016年11月4日起正式实施。这是继1992年《联合国气候变化框架公约》、1997年《京都议定书》之后,人类历史上应对气候变化的第三个里程碑式的国际法律文本,形成2020年后的全球气候治理格局。

《巴黎协定》共29条,包括目标、减缓、适应、损失损害、资金、技术、能力建设、透明度、全球盘点等内容。《巴黎协定》的长期目标是将全球平均气温较前工业化时期上升幅度控制在2℃以内,并努力将气温上升幅度限制在1.5℃以内。《巴黎协定》包括所有缔约方对减排和共同努力适应气候变化的承诺,并呼吁各国逐步加强承诺。《巴黎协定》为发达国家提供了协助发展中国家减缓和适应气候变化的方法,同时建立了透明监测和报告各国气候目标的框架。欧美等发达国家继续率先减排并开展绝对量化减排,为发展中国家提供资金支持;中国、印度等发展中国家应根据自身情况提高减排目标,逐步实现绝对减排或者限排目标;最不发达国家和小岛屿发展中国家可编制和通报反映它们特殊情况的关于温室气体排放发展的战略、计划和行动。

《巴黎协定》充分体现了联合国框架下各方的诉求,是一个非常平衡的协定。《巴黎协议》体现共同但有区别的责任原则,同时根据各自的国情和能力自主行动,采取非侵入、非对抗模式的评价机制,是一份让所有缔约国达成共识且都能参与的协议,有助于国际间(双边、多边机制)的合作和全球应对气候变化意识的培养。目前,共有189个国家加入了《巴黎协定》。

Paris Agreement

The Parties to this Agreement,

Being Parties to the *United Nations Framework Convention on Climate Change*, hereinafter referred to as "the *Convention*",

Pursuant to the Durban Platform for Enhanced Action established by decision 1/CP.17 of the Conference of the Parties to the *Convention* at its seventeenth session,

In pursuit of the objective of the *Convention*, and being guided by its principles, including the principle of equity and common but differentiated responsibilities and respective capabilities, in the light of different national circumstances,

Recognizing the need for an effective and progressive response to the urgent threat of climate change on the basis of the best available scientific knowledge,

Also recognizing the specific needs and special circumstances of developing country Parties, especially those that are particularly vulnerable to the adverse effects of climate change, as provided for in the *Convention*,

Taking full account of the specific needs and special situations of the least developed countries with regard to funding and transfer of technology,

Recognizing that Parties may be affected not only by climate change, but also by the impacts of the measures taken in response to it,

Emphasizing the intrinsic relationship that climate change actions, responses and impacts have with equitable access to sustainable development and eradication of poverty,

Recognizing the fundamental priority of safeguarding food security and ending hunger, and the particular vulnerabilities of food production systems to the adverse impacts of climate change,

Taking into account the imperatives of a just transition of the workforce and the creation of decent work and quality jobs in accordance with nationally defined development priorities,

Acknowledging that climate change is a common concern of humankind, Parties should, when taking action to address climate change, respect, promote and consider their respective obligations on human rights, the right to health, the rights of indigenous peoples, local communities, migrants, children, persons with disabilities and people in vulnerable situations and the right to development, as well as gender equality, empowerment of women and intergenerational equity,

Recognizing the importance of the conservation and enhancement, as appropriate, of sinks and reservoirs of the greenhouse gases referred to in the *Convention*,

Noting the importance of ensuring the integrity of all ecosystems, including oceans, and the protection of biodiversity, recognized by some cultures as Mother Earth, and noting the importance for some of the concept of "climate justice", when taking action to address climate change,

Affirming the importance of education, training, public awareness, public participation, public access to information and cooperation at all levels on the matters addressed in this Agreement,

Recognizing the importance of the engagements of all levels of government and various actors, in accordance with respective national legislations of Parties, in addressing climate change,

Also recognizing that sustainable lifestyles and sustainable patterns of consumption and production, with developed country Parties taking the lead, play an important role in addressing climate change,

Have agreed as follows:

Article 1

For the purpose of this Agreement, the definitions contained in Article 1 of the *Convention* shall apply. In addition:

1. "*Convention*" means the *United Nations Framework Convention on Climate Change*, adopted in New York on 9 May 1992.

2. "Conference of the Parties" means the Conference of the Parties to the *Convention*.

3. "Party" means a Party to this Agreement.

Article 2

1. This Agreement, in enhancing the implementation of the *Convention*, including its objective, aims to strengthen the global response to the threat of climate change, in the context of sustainable development and efforts to eradicate poverty, including by:

(a) Holding the increase in the global average temperature to well below 2℃ above pre-industrial levels and to pursue efforts to limit the temperature increase to 1.5℃ above pre-industrial levels, recognizing that this would significantly reduce the risks and impacts of climate change;

(b) Increasing the ability to adapt to the adverse impacts of climate change and foster climate resilience and low greenhouse gas emissions development, in a manner that does not threaten food production;

(c) Making finance flows consistent with a pathway towards low greenhouse gas emissions and climate-resilient development.

2. This Agreement will be implemented to reflect equity and the principle of common but differentiated responsibilities and respective capabilities, in the light of different national circumstances.

Article 3

As nationally determined contributions to the global response to climate change, all Parties are to undertake and communicate ambitious efforts as defined in Articles 4, 7, 9, 10, 11 and 13 with the view to achieving the purpose of this Agreement as set out in Article 2. The efforts of all Parties will represent a progression over time, while recognizing the need to support developing country Parties for the effective implementation of this Agreement.

Article 4

1. In order to achieve the long-term temperature goal set out in Article 2, Parties aim to reach global peaking of greenhouse gas emissions as soon as possible, recognizing that peaking will take longer for developing country Parties, and to undertake rapid reductions thereafter in accordance with best available science, so as to achieve a balance between anthropogenic emissions by sources and removals by sinks of greenhouse gases in the second half of this century, on the basis of equity, and in the context of sustainable development and efforts to eradicate poverty.

2. Each Party shall prepare, communicate and maintain successive nationally determined contributions that it intends to achieve. Parties shall pursue domestic mitigation measures, with the aim of achieving the objectives of such contributions.

3. Each Party's successive nationally determined contribution will represent a progression beyond the Party's then current nationally determined contribution and reflect its highest possible ambition, reflecting its common but differentiated responsibilities and respective capabilities, in the light of different national circumstances.

4. Developed country Parties should continue taking the lead by undertaking economy-wide absolute emission reduction targets. Developing country Parties should continue enhancing their mitigation efforts, and are encouraged to move over time towards economy-wide emission reduction or limitation targets in the light of different national circumstances.

5. Support shall be provided to developing country Parties for the implementation of this Article, in accordance with Articles 9, 10 and 11, recognizing that enhanced support for developing country Parties will allow for higher ambition in their actions.

6. The least developed countries and small island developing States may prepare and communicate strategies, plans and actions for low greenhouse gas emissions development reflecting their special circumstances.

7. Mitigation co-benefits resulting from Parties' adaptation actions and/or economic diversification plans can contribute to mitigation outcomes under this Article.

8. In communicating their nationally determined contributions, all Parties shall provide the information necessary for clarity, transparency and understanding in accordance with decision 1/CP.21 and any relevant decisions of the Conference of the Parties serving as the meeting of the Parties to the *Paris Agreement*.

9. Each Party shall communicate a nationally determined contribution every five years in accordance with decision 1/CP.21 and any relevant decisions of the Conference of the Parties serving as the meeting of the Parties to the *Paris Agreement* and be informed by the outcomes of the global stocktake referred to in Article 14.

10. The Conference of the Parties serving as the meeting of the Parties to the *Paris Agreement* shall consider common time frames for nationally determined contributions at its first session.

11. A Party may at any time adjust its existing nationally determined contribution with a view to enhancing its level of ambition, in accordance with guidance adopted by the Conference of the Parties serving as the meeting of the Parties to the *Paris Agreement*.

12. Nationally determined contributions communicated by Parties shall be recorded in a public registry maintained by the secretariat.

13. Parties shall account for their nationally determined contributions. In accounting for anthropogenic emissions and removals corresponding to their nationally determined contributions, Parties shall promote environmental integrity, transparency, accuracy, completeness, comparability and consistency, and ensure the avoidance of double counting, in accordance with guidance adopted by the Conference of the Parties serving as the meeting of the Parties to the *Paris Agreement*.

14. In the context of their nationally determined contributions, when recognizing and implementing mitigation actions with respect to anthropogenic emissions and removals, Parties should take into account, as appropriate, existing methods and guidance under the *Convention*, in the light of the provisions of paragraph 13 of this Article.

15. Parties shall take into consideration in the implementation of this Agreement the concerns of Parties with economies most affected by the impacts of response measures, particularly developing country Parties.

16. Parties, including regional economic integration organizations and their member States, that have reached an agreement to act jointly under paragraph 2 of this Article shall notify the secretariat of the terms of that agreement, including the emission level allocated to each Party within the relevant time period, when they communicate their nationally determined contributions. The secretariat shall in turn inform the Parties and signatories to the *Convention* of the terms of that agreement.

17. Each party to such an agreement shall be responsible for its emission level as set out in the agreement referred to in paragraph 16 above in accordance with paragraphs 13 and 14 of this Article and Articles 13 and 15.

18. If Parties acting jointly do so in the framework of, and together with, a regional economic integration organization which is itself a Party to this Agreement, each member State of that regional economic integration organization individually, and together with the regional economic integration organization, shall be responsible for its emission level as set out in the agreement communicated under paragraph 16 of this Article in accordance with paragraphs 13 and 14 of this Article and Articles 13 and 15.

19. All Parties should strive to formulate and communicate long-term low greenhouse gas emission development strategies, mindful of Article 2 taking into account their common but differentiated responsibilities and respective capabilities, in the light of different national circumstances.

Article 5

1. Parties should take action to conserve and enhance, as appropriate, sinks and reservoirs of greenhouse gases as referred to in Article 4, paragraph 1(d), of the *Convention*, including

forests.

2. Parties are encouraged to take action to implement and support, including through results-based payments, the existing framework as set out in related guidance and decisions already agreed under the *Convention* for: policy approaches and positive incentives for activities relating to reducing emissions from deforestation and forest degradation, and the role of conservation, sustainable management of forests and enhancement of forest carbon stocks in developing countries; and alternative policy approaches, such as joint mitigation and adaptation approaches for the integral and sustainable management of forests, while reaffirming the importance of incentivizing, as appropriate, non-carbon benefits associated with such approaches.

Article 6

1. Parties recognize that some Parties choose to pursue voluntary cooperation in the implementation of their nationally determined contributions to allow for higher ambition in their mitigation and adaptation actions and to promote sustainable development and environmental integrity.

2. Parties shall, where engaging on a voluntary basis in cooperative approaches that involve the use of internationally transferred mitigation outcomes towards nationally determined contributions, promote sustainable development and ensure environmental integrity and transparency, including in governance, and shall apply robust accounting to ensure, inter alia, the avoidance of double counting, consistent with guidance adopted by the Conference of the Parties serving as the meeting of the Parties to the *Paris Agreement*.

3. The use of internationally transferred mitigation outcomes to achieve nationally determined contributions under this Agreement shall be voluntary and authorized by participating Parties.

4. A mechanism to contribute to the mitigation of greenhouse gas emissions and support sustainable development is hereby established under the authority and guidance of the Conference of the Parties serving as the meeting of the Parties to the *Paris Agreement* for use by Parties on a voluntary basis. It shall be supervised by a body designated by the Conference of the Parties serving as the meeting of the Parties to the *Paris Agreement*, and shall aim:

(a) To promote the mitigation of greenhouse gas emissions while fostering sustainable development;

(b) To incentivize and facilitate participation in the mitigation of greenhouse gas emissions by public and private entities authorized by a Party;

(c) To contribute to the reduction of emission levels in the host Party, which will benefit from mitigation activities resulting in emission reductions that can also be used by another Party to fulfill its nationally determined contribution; and

(d) To deliver an overall mitigation in global emissions.

5. Emission reductions resulting from the mechanism referred to in paragraph 4 of this Article shall not be used to demonstrate achievement of the host Party's nationally determined contribution if used by another Party to demonstrate achievement of its nationally determined

contribution.

6. The Conference of the Parties serving as the meeting of the Parties to the *Paris Agreement* shall ensure that a share of the proceeds from activities under the mechanism referred to in paragraph 4 of this Article is used to cover administrative expenses as well as to assist developing country Parties that are particularly vulnerable to the adverse effects of climate change to meet the costs of adaptation.

7. The Conference of the Parties serving as the meeting of the Parties to the *Paris Agreement* shall adopt rules, modalities and procedures for the mechanism referred to in paragraph 4 of this Article at its first session.

8. Parties recognize the importance of integrated, holistic and balanced non-market approaches being available to Parties to assist in the implementation of their nationally determined contributions, in the context of sustainable development and poverty eradication, in a coordinated and effective manner, including through, inter alia, mitigation, adaptation, finance, technology transfer and capacity-building, as appropriate. These approaches shall aim to:

(a) Promote mitigation and adaptation ambition;

(b) Enhance public and private sector participation in the implementation of nationally determined contributions; and

(c) Enable opportunities for coordination across instruments and relevant institutional arrangements.

9. A framework for non-market approaches to sustainable development is hereby defined to promote the non-market approaches referred to in paragraph 8 of this Article.

Article 7

1. Parties hereby establish the global goal on adaptation of enhancing adaptive capacity, strengthening resilience and reducing vulnerability to climate change, with a view to contributing to sustainable development and ensuring an adequate adaptation response in the context of the temperature goal referred to in Article 2.

2. Parties recognize that adaptation is a global challenge faced by all with local, subnational, national, regional and international dimensions, and that it is a key component of and makes a contribution to the long-term global response to climate change to protect people, livelihoods and ecosystems, taking into account the urgent and immediate needs of those developing country Parties that are particularly vulnerable to the adverse effects of climate change.

3. The adaptation efforts of developing country Parties shall be recognized, in accordance with the modalities to be adopted by the Conference of the Parties serving as the meeting of the Parties to the *Paris Agreement* at its first session.

4. Parties recognize that the current need for adaptation is significant and that greater levels of mitigation can reduce the need for additional adaptation efforts, and that greater adaptation needs can involve greater adaptation costs.

5. Parties acknowledge that adaptation action should follow a country-driven, gender-responsive,

participatory and fully transparent approach, taking into consideration vulnerable groups, communities and ecosystems, and should be based on and guided by the best available science and, as appropriate, traditional knowledge, knowledge of indigenous peoples and local knowledge systems, with a view to integrating adaptation into relevant socioeconomic and environmental policies and actions, where appropriate.

6. Parties recognize the importance of support for and international cooperation on adaptation efforts and the importance of taking into account the needs of developing country Parties, especially those that are particularly vulnerable to the adverse effects of climate change.

7. Parties should strengthen their cooperation on enhancing action on adaptation, taking into account the *Cancun Adaptation Framework*, including with regard to:

(a) Sharing information, good practices, experiences and lessons learned, including, as appropriate, as these relate to science, planning, policies and implementation in relation to adaptation actions;

(b) Strengthening institutional arrangements, including those under the *Convention* that serve this Agreement, to support the synthesis of relevant information and knowledge, and the provision of technical support and guidance to Parties;

(c) Strengthening scientific knowledge on climate, including research, systematic observation of the climate system and early warning systems, in a manner that informs climate services and supports decision-making;

(d) Assisting developing country Parties in identifying effective adaptation practices, adaptation needs, priorities, support provided and received for adaptation actions and efforts, and challenges and gaps, in a manner consistent with encouraging good practices;

(e) Improving the effectiveness and durability of adaptation actions.

8. United Nations specialized organizations and agencies are encouraged to support the efforts of Parties to implement the actions referred to in paragraph 7 of this Article, taking into account the provisions of paragraph 5 of this Article.

9. Each Party shall, as appropriate, engage in adaptation planning processes and the implementation of actions, including the development or enhancement of relevant plans, policies and/or contributions, which may include:

(a) The implementation of adaptation actions, undertakings and/or efforts;

(b) The process to formulate and implement national adaptation plans;

(c) The assessment of climate change impacts and vulnerability, with a view to formulating nationally determined prioritized actions, taking into account vulnerable people, places and ecosystems;

(d) Monitoring and evaluating and learning from adaptation plans, policies, programmes and actions; and

(e) Building the resilience of socioeconomic and ecological systems, including through economic diversification and sustainable management of natural resources.

10. Each Party should, as appropriate, submit and update periodically an adaptation

communication, which may include its priorities, implementation and support needs, plans and actions, without creating any additional burden for developing country Parties.

11. The adaptation communication referred to in paragraph 10 of this Article shall be, as appropriate, submitted and updated periodically, as a component of or in conjunction with other communications or documents, including a national adaptation plan, a nationally determined contribution as referred to in Article 4, paragraph 2, and/or a national communication.

12. The adaptation communications referred to in paragraph 10 of this Article shall be recorded in a public registry maintained by the secretariat.

13. Continuous and enhanced international support shall be provided to developing country Parties for the implementation of paragraphs 7, 9, 10 and 11 of this Article, in accordance with the provisions of Articles 9, 10 and 11.

14. The global stocktake referred to in Article 14 shall, inter alia:

(a) Recognize adaptation efforts of developing country Parties;

(b) Enhance the implementation of adaptation action taking into account the adaptation communication referred to in paragraph 10 of this Article;

(c) Review the adequacy and effectiveness of adaptation and support provided for adaptation; and

(d) Review the overall progress made in achieving the global goal on adaptation referred to in paragraph 1 of this Article.

Article 8

1. Parties recognize the importance of averting, minimizing and addressing loss and damage associated with the adverse effects of climate change, including extreme weather events and slow onset events, and the role of sustainable development in reducing the risk of loss and damage.

2. The Warsaw International Mechanism for Loss and Damage associated with Climate Change Impacts shall be subject to the authority and guidance of the Conference of the Parties serving as the meeting of the Parties to the *Paris Agreement* and may be enhanced and strengthened, as determined by the Conference of the Parties serving as the meeting of the Parties to the *Paris Agreement*.

3. Parties should enhance understanding, action and support, including through the Warsaw International Mechanism, as appropriate, on a cooperative and facilitative basis with respect to loss and damage associated with the adverse effects of climate change.

4. Accordingly, areas of cooperation and facilitation to enhance understanding, action and support may include:

(a) Early warning systems;

(b) Emergency preparedness;

(c) Slow onset events;

(d) Events that may involve irreversible and permanent loss and damage;

(e) Comprehensive risk assessment and management;

(f) Risk insurance facilities, climate risk pooling and other insurance solutions;

(g) Non-economic losses;

(h) Resilience of communities, livelihoods and ecosystems.

5. The Warsaw International Mechanism shall collaborate with existing bodies and expert groups under the Agreement, as well as relevant organizations and expert bodies outside the Agreement.

Article 9

1. Developed country Parties shall provide financial resources to assist developing country Parties with respect to both mitigation and adaptation in continuation of their existing obligations under the *Convention*.

2. Other Parties are encouraged to provide or continue to provide such support voluntarily.

3. As part of a global effort, developed country Parties should continue to take the lead in mobilizing climate finance from a wide variety of sources, instruments and channels, noting the significant role of public funds, through a variety of actions, including supporting country-driven strategies, and taking into account the needs and priorities of developing country Parties. Such mobilization of climate finance should represent a progression beyond previous efforts.

4. The provision of scaled-up financial resources should aim to achieve a balance between adaptation and mitigation, taking into account country-driven strategies, and the priorities and needs of developing country Parties, especially those that are particularly vulnerable to the adverse effects of climate change and have significant capacity constraints, such as the least developed countries and small island developing States, considering the need for public and grant-based resources for adaptation.

5. Developed country Parties shall biennially communicate indicative quantitative and qualitative information related to paragraphs 1 and 3 of this Article, as applicable, including, as available, projected levels of public financial resources to be provided to developing country Parties. Other Parties providing resources are encouraged to communicate biennially such information on a voluntary basis.

6. The global stocktake referred to in Article 14 shall take into account the relevant information provided by developed country Parties and/or Agreement bodies on efforts related to climate finance.

7. Developed country Parties shall provide transparent and consistent information on support for developing country Parties provided and mobilized through public interventions biennially in accordance with the modalities, procedures and guidelines to be adopted by the Conference of the Parties serving as the meeting of the Parties to the *Paris Agreement*, at its first session, as stipulated in Article 13, paragraph 13. Other Parties are encouraged to do so.

8. The Financial Mechanism of the *Convention*, including its operating entities, shall serve as the financial mechanism of this Agreement.

9. The institutions serving this Agreement, including the operating entities of the Financial Mechanism of the *Convention*, shall aim to ensure efficient access to financial

resources through simplified approval procedures and enhanced readiness support for developing country Parties, in particular for the least developed countries and small island developing States, in the context of their national climate strategies and plans.

Article 10

1. Parties share a long-term vision on the importance of fully realizing technology development and transfer in order to improve resilience to climate change and to reduce greenhouse gas emissions.

2. Parties, noting the importance of technology for the implementation of mitigation and adaptation actions under this Agreement and recognizing existing technology deployment and dissemination efforts, shall strengthen cooperative action on technology development and transfer.

3. The Technology Mechanism established under the *Convention* shall serve this Agreement.

4. A technology framework is hereby established to provide overarching guidance to the work of the Technology Mechanism in promoting and facilitating enhanced action on technology development and transfer in order to support the implementation of this Agreement, in pursuit of the long-term vision referred to in paragraph 1 of this Article.

5. Accelerating, encouraging and enabling innovation is critical for an effective, long-term global response to climate change and promoting economic growth and sustainable development. Such effort shall be, as appropriate, supported, including by the Technology Mechanism and, through financial means, by the Financial Mechanism of the *Convention*, for collaborative approaches to research and development, and facilitating access to technology, in particular for early stages of the technology cycle, to developing country Parties.

6. Support, including financial support, shall be provided to developing country Parties for the implementation of this Article, including for strengthening cooperative action on technology development and transfer at different stages of the technology cycle, with a view to achieving a balance between support for mitigation and adaptation. The global stocktake referred to in Article 14 shall take into account available information on efforts related to support on technology development and transfer for developing country Parties.

Article 11

1. Capacity-building under this Agreement should enhance the capacity and ability of developing country Parties, in particular countries with the least capacity, such as the least developed countries, and those that are particularly vulnerable to the adverse effects of climate change, such as small island developing States, to take effective climate change action, including, inter alia, to implement adaptation and mitigation actions, and should facilitate technology development, dissemination and deployment, access to climate finance, relevant aspects of education, training and public awareness, and the transparent, timely and accurate communication of information.

2. Capacity-building should be country-driven, based on and responsive to national needs, and foster country ownership of Parties, in particular, for developing country Parties,

including at the national, subnational and local levels. Capacity-building should be guided by lessons learned, including those from capacity-building activities under the *Convention*, and should be an effective, iterative process that is participatory, cross-cutting and gender-responsive.

3. All Parties should cooperate to enhance the capacity of developing country Parties to implement this Agreement. Developed country Parties should enhance support for capacity-building actions in developing country Parties.

4. All Parties enhancing the capacity of developing country Parties to implement this Agreement, including through regional, bilateral and multilateral approaches, shall regularly communicate on these actions or measures on capacity-building. Developing country Parties should regularly communicate progress made on implementing capacity-building plans, policies, actions or measures to implement this Agreement.

5. Capacity-building activities shall be enhanced through appropriate institutional arrangements to support the implementation of this Agreement, including the appropriate institutional arrangements established under the *Convention* that serve this Agreement. The Conference of the Parties serving as the meeting of the Parties to the *Paris Agreement* shall, at its first session, consider and adopt a decision on the initial institutional arrangements for capacity-building.

Article 12

Parties shall cooperate in taking measures, as appropriate, to enhance climate change education, training, public awareness, public participation and public access to information, recognizing the importance of these steps with respect to enhancing actions under this Agreement.

Article 13

1. In order to build mutual trust and confidence and to promote effective implementation, an enhanced transparency framework for action and support, with built-in flexibility which takes into account Parties' different capacities and builds upon collective experience is hereby established.

2. The transparency framework shall provide flexibility in the implementation of the provisions of this Article to those developing country Parties that need it in the light of their capacities. The modalities, procedures and guidelines referred to in paragraph 13 of this Article shall reflect such flexibility.

3. The transparency framework shall build on and enhance the transparency arrangements under the *Convention*, recognizing the special circumstances of the least developed countries and small island developing States, and be implemented in a facilitative, non-intrusive, non-punitive manner, respectful of national sovereignty, and avoid placing undue burden on Parties.

4. The transparency arrangements under the *Convention*, including national communications, biennial reports and biennial update reports, international assessment and review and international consultation and analysis, shall form part of the experience drawn

upon for the development of the modalities, procedures and guidelines under paragraph 13 of this Article.

5. The purpose of the framework for transparency of action is to provide a clear understanding of climate change action in the light of the objective of the *Convention* as set out in its Article 2, including clarity and tracking of progress towards achieving Parties' individual nationally determined contributions under Article 4, and Parties' adaptation actions under Article 7, including good practices, priorities, needs and gaps, to inform the global stocktake under Article 14.

6. The purpose of the framework for transparency of support is to provide clarity on support provided and received by relevant individual Parties in the context of climate change actions under Articles 4, 7, 9, 10 and 11, and, to the extent possible, to provide a full overview of aggregate financial support provided, to inform the global stocktake under Article 14.

7. Each Party shall regularly provide the following information:

(a) A national inventory report of anthropogenic emissions by sources and removals by sinks of greenhouse gases, prepared using good practice methodologies accepted by the Intergovernmental Panel on Climate Change and agreed upon by the Conference of the Parties serving as the meeting of the Parties to the *Paris Agreement*;

(b) Information necessary to track progress made in implementing and achieving its nationally determined contribution under Article 4.

8. Each Party should also provide information related to climate change impacts and adaptation under Article 7, as appropriate.

9. Developed country Parties shall, and other Parties that provide support should, provide information on financial, technology transfer and capacity-building support provided to developing country Parties under Article 9, 10 and 11.

10. Developing country Parties should provide information on financial, technology transfer and capacity-building support needed and received under Articles 9, 10 and 11.

11. Information submitted by each Party under paragraphs 7 and 9 of this Article shall undergo a technical expert review, in accordance with decision 1/CP.21. For those developing country Parties that need it in the light of their capacities, the review process shall include assistance in identifying capacity-building needs. In addition, each Party shall participate in a facilitative, multilateral consideration of progress with respect to efforts under Article 9, and its respective implementation and achievement of its nationally determined contribution.

12. The technical expert review under this paragraph shall consist of a consideration of the Party's support provided, as relevant, and its implementation and achievement of its nationally determined contribution. The review shall also identify areas of improvement for the Party, and include a review of the consistency of the information with the modalities, procedures and guidelines referred to in paragraph 13 of this Article, taking into account the flexibility accorded to the Party under paragraph 2 of this Article. The review shall pay particular attention to the respective national capabilities and circumstances of developing country Parties.

13. The Conference of the Parties serving as the meeting of the Parties to the *Paris Agreement* shall, at its first session, building on experience from the arrangements related to transparency under the *Convention*, and elaborating on the provisions in this Article, adopt common modalities, procedures and guidelines, as appropriate, for the transparency of action and support.

14. Support shall be provided to developing countries for the implementation of this Article.

15. Support shall also be provided for the building of transparency-related capacity of developing country Parties on a continuous basis.

Article 14

1. The Conference of the Parties serving as the meeting of the Parties to the *Paris Agreement* shall periodically take stock of the implementation of this Agreement to assess the collective progress towards achieving the purpose of this Agreement and its long-term goals (referred to as the "global stocktake"). It shall do so in a comprehensive and facilitative manner, considering mitigation, adaptation and the means of implementation and support, and in the light of equity and the best available science.

2. The Conference of the Parties serving as the meeting of the Parties to the *Paris Agreement* shall undertake its first global stocktake in 2023 and every five years thereafter unless otherwise decided by the Conference of the Parties serving as the meeting of the Parties to the *Paris Agreement*.

3. The outcome of the global stocktake shall inform Parties in updating and enhancing, in a nationally determined manner, their actions and support in accordance with the relevant provisions of this Agreement, as well as in enhancing international cooperation for climate action.

Article 15

1. A mechanism to facilitate implementation of and promote compliance with the provisions of this Agreement is hereby established.

2. The mechanism referred to in paragraph 1 of this Article shall consist of a committee that shall be expert-based and facilitative in nature and function in a manner that is transparent, non-adversarial and non-punitive. The committee shall pay particular attention to the respective national capabilities and circumstances of Parties.

3. The committee shall operate under the modalities and procedures adopted by the Conference of the Parties serving as the meeting of the Parties to the *Paris Agreement* at its first session and report annually to the Conference of the Parties serving as the meeting of the Parties to the *Paris Agreement*.

Article 16

1. The Conference of the Parties, the supreme body of the *Convention*, shall serve as the meeting of the Parties to this Agreement.

2. Parties to the *Convention* that are not Parties to this Agreement may participate as observers in the proceedings of any session of the Conference of the Parties serving as the

meeting of the Parties to this Agreement. When the Conference of the Parties serves as the meeting of the Parties to this Agreement, decisions under this Agreement shall be taken only by those that are Parties to this Agreement.

3. When the Conference of the Parties serves as the meeting of the Parties to this Agreement, any member of the Bureau of the Conference of the Parties representing a Party to the *Convention* but, at that time, not a Party to this Agreement, shall be replaced by an additional member to be elected by and from amongst the Parties to this Agreement.

4. The Conference of the Parties serving as the meeting of the Parties to the *Paris Agreement* shall keep under regular review the implementation of this Agreement and shall make, within its mandate, the decisions necessary to promote its effective implementation. It shall perform the functions assigned to it by this Agreement and shall:

(a) Establish such subsidiary bodies as deemed necessary for the implementation of this Agreement; and

(b) Exercise such other functions as may be required for the implementation of this Agreement.

5. The rules of procedure of the Conference of the Parties and the financial procedures applied under the *Convention* shall be applied mutatis mutandis under this Agreement, except as may be otherwise decided by consensus by the Conference of the Parties serving as the meeting of the Parties to the *Paris Agreement*.

6. The first session of the Conference of the Parties serving as the meeting of the Parties to the *Paris Agreement* shall be convened by the secretariat in conjunction with the first session of the Conference of the Parties that is scheduled after the date of entry into force of this Agreement. Subsequent ordinary sessions of the Conference of the Parties serving as the meeting of the Parties to the *Paris Agreement* shall be held in conjunction with ordinary sessions of the Conference of the Parties, unless otherwise decided by the Conference of the Parties serving as the meeting of the Parties to the *Paris Agreement*.

7. Extraordinary sessions of the Conference of the Parties serving as the meeting of the Parties to the *Paris Agreement* shall be held at such other times as may be deemed necessary by the Conference of the Parties serving as the meeting of the Parties to the *Paris Agreement* or at the written request of any Party, provided that, within six months of the request being communicated to the Parties by the secretariat, it is supported by at least one third of the Parties.

8. The United Nations and its specialized agencies and the International Atomic Energy Agency, as well as any State member thereof or observers thereto not party to the *Convention*, may be represented at sessions of the Conference of the Parties serving as the meeting of the Parties to the *Paris Agreement* as observers. Any body or agency, whether national or international, governmental or non-governmental, which is qualified in matters covered by this Agreement and which has informed the secretariat of its wish to be represented at a session of the Conference of the Parties serving as the meeting of the Parties to the *Paris Agreement* as an observer, may be so admitted unless at least one third of the Parties present object. The admission and participation of observers shall be subject to the rules of procedure

referred to in paragraph 5 of this Article.

Article 17

1. The secretariat established by Article 8 of the *Convention* shall serve as the secretariat of this Agreement.

2. Article 8, paragraph 2, of the *Convention* on the functions of the secretariat, and Article 8, paragraph 3, of the *Convention*, on the arrangements made for the functioning of the secretariat, shall apply mutatis mutandis to this Agreement. The secretariat shall, in addition, exercise the functions assigned to it under this Agreement and by the Conference of the Parties serving as the meeting of the Parties to the *Paris Agreement*.

Article 18

1. The Subsidiary Body for Scientific and Technological Advice and the Subsidiary Body for Implementation established by Articles 9 and 10 of the *Convention* shall serve, respectively, as the Subsidiary Body for Scientific and Technological Advice and the Subsidiary Body for Implementation of this Agreement. The provisions of the *Convention* relating to the functioning of these two bodies shall apply mutatis mutandis to this Agreement. Sessions of the meetings of the Subsidiary Body for Scientific and Technological Advice and the Subsidiary Body for Implementation of this Agreement shall be held in conjunction with the meetings of, respectively, the Subsidiary Body for Scientific and Technological Advice and the Subsidiary Body for Implementation of the *Convention*.

2. Parties to the *Convention* that are not Parties to this Agreement may participate as observers in the proceedings of any session of the subsidiary bodies. When the subsidiary bodies serve as the subsidiary bodies of this Agreement, decisions under this Agreement shall be taken only by those that are Parties to this Agreement.

3. When the subsidiary bodies established by Articles 9 and 10 of the *Convention* exercise their functions with regard to matters concerning this Agreement, any member of the bureaux of those subsidiary bodies representing a Party to the Convention but, at that time, not a Party to this Agreement, shall be replaced by an additional member to be elected by and from amongst the Parties to this Agreement.

Article 19

1. Subsidiary bodies or other institutional arrangements established by or under the *Convention*, other than those referred to in this Agreement, shall serve this Agreement upon a decision of the Conference of the Parties serving as the meeting of the Parties to the *Paris Agreement*. The Conference of the Parties serving as the meeting of the Parties to the *Paris Agreement* shall specify the functions to be exercised by such subsidiary bodies or arrangements.

2. The Conference of the Parties serving as the meeting of the Parties to the *Paris Agreement* may provide further guidance to such subsidiary bodies and institutional arrangements.

Article 20

1. This Agreement shall be open for signature and subject to ratification, acceptance or

approval by States and regional economic integration organizations that are Parties to the Convention. It shall be open for signature at the United Nations Headquarters in New York from 22 April 2016 to 21 April 2017. Thereafter, this Agreement shall be open for accession from the day following the date on which it is closed for signature. Instruments of ratification, acceptance, approval or accession shall be deposited with the Depositary.

2. Any regional economic integration organization that becomes a Party to this Agreement without any of its member States being a Party shall be bound by all the obligations under this Agreement. In the case of regional economic integration organizations with one or more member States that are Parties to this Agreement, the organization and its member States shall decide on their respective responsibilities for the performance of their obligations under this Agreement. In such cases, the organization and the member States shall not be entitled to exercise rights under this Agreement concurrently.

3. In their instruments of ratification, acceptance, approval or accession, regional economic integration organizations shall declare the extent of their competence with respect to the matters governed by this Agreement. These organizations shall also inform the Depositary, who shall in turn inform the Parties, of any substantial modification in the extent of their competence.

Article 21

1. This Agreement shall enter into force on the thirtieth day after the date on which at least 55 Parties to the *Convention* accounting in total for at least an estimated 55% of the total global greenhouse gas emissions have deposited their instruments of ratification, acceptance, approval or accession.

2. Solely for the limited purpose of paragraph 1 of this Article, "total global greenhouse gas emissions" means the most up-to-date amount communicated on or before the date of adoption of this Agreement by the Parties to the *Convention*.

3. For each State or regional economic integration organization that ratifies, accepts or approves this Agreement or accedes thereto after the conditions set out in paragraph 1 of this Article for entry into force have been fulfilled, this Agreement shall enter into force on the thirtieth day after the date of deposit by such State or regional economic integration organization of its instrument of ratification, acceptance, approval or accession.

4. For the purposes of paragraph 1 of this Article, any instrument deposited by a regional economic integration organization shall not be counted as additional to those deposited by its member States.

Article 22

The provisions of Article 15 of the *Convention* on the adoption of amendments to the *Convention* shall apply mutatis mutandis to this Agreement.

Article 23

1. The provisions of Article 16 of the Convention on the adoption and amendment of annexes to the *Convention* shall apply mutatis mutandis to this Agreement.

2. Annexes to this Agreement shall form an integral part thereof and, unless otherwise

expressly provided for, a reference to this Agreement constitutes at the same time a reference to any annexes thereto. Such annexes shall be restricted to lists, forms and any other material of a descriptive nature that is of a scientific, technical, procedural or administrative character.

Article 24

The provisions of Article 14 of the *Convention* on settlement of disputes shall apply mutatis mutandis to this Agreement.

Article 25

1. Each Party shall have one vote, except as provided for paragraph 2 of this Article.

2. Regional economic integration organizations, in matters within their competence, shall exercise their right to vote with a number of votes equal to the number of their member States that are Parties to this Agreement. Such an organization shall not exercise its right to vote if any of its member States exercises its right, and vice versa.

Article 26

The Secretary-General of the United Nations shall be the Depositary of this Agreement.

Article 27

No reservations may be made to this Agreement.

Article 28

1. At any time after three years from the date on which this Agreement has entered into force for a Party, that Party may withdraw from this Agreement by giving written notification to the Depositary.

2. Any such withdrawal shall take effect upon expiry of one year from the date of receipt by the Depositary of the notification of withdrawal, or on such later date as may be specified in the notification of withdrawal.

3. Any Party that withdraws from the *Convention* shall be considered as also having withdrawn from this Agreement.

Article 29

The original of this Agreement, of which the Arabic, Chinese, English, French, Russian and Spanish texts are equally authentic, shall be deposited with the Secretary-General of the United Nations.

DONE at Paris this twelfth day of December two thousand and fifteen.

IN WITNESS WHEREOF, the undersigned, being duly authorized to that effect, have signed this Agreement.

Notes

1. Durban Platform for Enhanced Action

德班加强行动平台。2011 年南非德班气候大会通过了《京都议定书》工作组和《联合国气候变化框架公约》(简称《公约》)长期合作行动特设工作组的决议,建立了德班加强行动特设工作组,负责 2020 年后减排温室气体的具体安排。该工作组主要任务是制定一个适用于所有《公约》缔约方的法律工具,作为 2020 年后各方加强《公约》实施、减控温室气体排放和应对气候变化的依据。这项工作于 2012 年上半年开始，2015 年前结束。各缔约方要在工作组工作的基础上，从 2020 年开始根据该法律工具或者法律成果探讨如何减排，降低温室气体排放。

2. Regional Economic Integration

区域经济一体化，也称"区域经济集团化"，是世界经济发展的必然结果，伴随着经济全球化的推进而不断发展升级。自 20 世纪 50 年代末以来，一些地理位置相近的国家或地区间通过加强经济合作，为谋求风险成本和机会成本的最小化和利益的最大化，逐步让渡部分甚至全部经济主权，采取共同的经济政策，形成了一体化程度较高的区域经济合作组织或国家集团。其组织形式按一体化程度由低到高排列，包括优惠贸易安排、自由贸易区、关税同盟、共同市场、经济联盟和完全的经济一体化等。目前，一体化程度较高的著名区域经济集团有欧洲联盟、北美自由贸易区、亚太经济合作组织等。全球范围内的区域经济一体化迅速发展主要依靠 3 条途径：一是不断深化、升级现有形式；二是扩展现有集团成员；三是缔结新的区域贸易协议或重新启动沉寂多年的区域经济合作谈判。区域经济一体化覆盖大多数国家和地区。据世界银行统计，全球只有 12 个岛国和公国没有参与任何区域贸易协议。174 个国家和地区参加了至少一个（最多达 29 个）区域贸易协议，平均每个国家或地区参加了 5 个。区域经济一体化在内容上也得到广泛深入。新一轮的区域协议涵盖的范围大大扩展，不仅包括货物贸易自由化，而且包括服务贸易自由化、农产品贸易自由化、投资自由化、贸易争端解决机制、统一的竞争政策、知识产权保护标准、共同的环境标准、劳工标准，甚至提出要具备共同的民主理念等。

3. International Atomic Energy Agency

国际原子能机构（简称 IAEA），是一个同联合国建立关系，并由世界各国政府在原子能领域进行科学技术合作的机构，总部设在奥地利的维也纳。1954 年 12 月，第 9 届联合国大会通过决议，要求成立一个专门致力于和平利用原子能的国际机构。经过两年筹备，有 82 个国家参加的规约会议于 1956 年 10 月 26 日通过了《国际原子能机构规约》（以下简称《规约》）。1957 年 7 月 29 日，《规约》正式生效。同年 10 月，国际原子能机构召开首次全体会议，宣布机构正式成立。该机构的宗旨是谋求加速和扩大原子能对全世界和平、健康及繁荣的贡献，确保由其本身、或经其请求、或在其监督或管制下提供的援助不用于推进任何军事目的。国际原子能机构的组织机构包括大会、理事会和秘书处。大会由全体成员国代表参加，每年召开一次。秘书处由总干事领导下的专业人员和工作人员组成，下设总干事办公室、管理司、核科学核应用司、保障司、技术合作司、核能司、核安全与核安保司。总干事由理事会任命，6 名副总干事负责 6 个独立的部门。理事会由 35 国组成，为该组织最高执行机构；下设科学咨询委员会、技术援助委员会、行政和预算委员会和保障委员会。

4. the United Nations Framework Convention on Climate Change

《联合国气候变化框架公约》（简称《公约》），由联合国大会于 1992 年 5 月 9 日通过，1992 年 6 月在巴西里约热内卢召开的由世界各国政府首脑参加的联合国环境与发展会议期间开放签署，在地球峰会上由 150 多个国家以及欧洲经济共同体共同签署。《公约》由序言及 26 条正文组成，具有法律约束力，终极目标是减少温室气体排放，减少人为活动对气候系统的危害，减缓气候变化，增强生态系统对气候变化的适应性，确保粮食生产和经济可持续发展。根据"共同但有区别的责任"原则，《公约》对发达国家和发展中国家规定的义务以及履行义务的程序有所区别，要求发达国家作为温室气体的排放大户，采取具体措施限制温室气体的排放，并向发展中国家提供资金以支付他们履行公约义务所需的费用。而发展中国家只承担提供温室气体源与温室气体汇的国家清单的义务，制订并执行含有关于温室气体源与汇方面措施的方案，不承担有法律约束力的限控义务。《公约》建立了一个向发展中国家提供资金和技术，使其能够履行公约义务的机制。《公约》是世界上第一个为全面控制二氧化碳等温室气体排放，应对全球气候变暖给人类经济和社会带来不利影响的国际公约，也是国际社会在应对全球气候变化问题上进行国际合作的一个基本框架。据统计，如今已有190多个国家批准了《公约》，这些国家被称为《公约》缔约方。《公约》缔约方作出了许多旨在解决气候变化问题的承诺。每个缔约方都必须定期提交专项报告，其内容必须包含该缔约方的温室气体排放信息，并说明为实施《公约》所执行的计划及具体措施。《公约》于1994年3月生效，奠定了应对气候变化国际合作的法律基础，是具有权威性、普遍性、全面性的国际框架。

5. the Warsaw International Mechanism for Loss and Damage

华沙损失与损害国际机制。2013年11月，《联合国气候变化框架公约》缔约方会议第19届会议设立了与气候变化不利影响相关的华沙损失与损害国际机制，去处理特别易受气候变化不利影响的发展中国家与气候变化影响相关的损失和损害问题，包括极端事件和缓发事件的影响。华沙损失与损害国际机制的职能包括：①增进对处理与气候变化不利影响(包括缓发事件影响)相关的损失和损害的全面风险管理办法的认识和了解；②加强相关利害关系方之间的对话、协调、统一和协同；③加强行动和支持，包括资金、技术和能力建设方面的行动和支持，以处理与气候变化不利影响相关的损失和损害。

Key Words and Phrases

1. amendment	/əˈmendmənt/	n.	the process of changing a law or a document（法律、文件的）改动，修正案，修改，修订
2. anthropogenic	/ˌænθrəpəˈdʒenɪk/	adj.	of or relating to the study of the origins and development of human beings 人为的，人类活动产生的
3. authorize	/ˈɔːθəraɪz/	v.	to give official permission for sth., or for sb. to do sth. 批准；授权
4. biennial	/baɪˈeniəl/	adj.	happening once every two years 两年一次的
5. collaborative	/kəˈlæbərətɪv/	adj.	(formal) involving, or done by, several people or groups of people working together 合作的；协作的；协力的
6. comprehensive	/ˌkɒmprɪˈhensɪv/	adj.	including all, or almost all, the items, details, facts, information, etc. that may

			be concerned 综合性的，全部的；详尽的
7. convene	/kən'vi:n/	v.	to arrange for people to come together for a formal meeting 召集，召开（正式会议）
8. deployment	/dɪ'plɔɪmənt/	n.	to organize, move resources, troops or equipment so that they are ready for quick action; (formal) to use sth. effectively 部署，调动，调集；有效利用
9. depositary	/dɪ'pɒzɪt(ə)ri/	n.	a person or a facility where things can be deposited for storage or safekeeping eradication 保管人；受托人；受托公司；存放处
10. eradication	/ɪˌrædɪ'keɪʃ(ə)n/	n.	the complete destruction of every trace of something 根除，消灭，杜绝
11. facilitative	/fə'sɪlɪˌteɪtɪv/	adj.	freeing from difficulty or impediment 提供便利的，使顺利的，促进的
12. hereinafter	/ˌhɪərɪn'ɑːftə(r)/	adv.	in a subsequent part of this document or statement or matter etc. （文件、声明等中）在下文中
13. imperative	/ɪm'perətɪv/	n.	(formal) a thing that is very important and needs immediate attention or action 重要紧急的事；必要的事
		adj.	(formal) very important and needing immediate attention or action 重要紧急的；迫切的
14. incentivize	/ɪn'sentɪvaɪz/	v.	to encourage sb. to behave in a particular way by offering them a reward 激励，奖励
15. intrinsic	/ɪn'trɪnzɪk/	adj.	belonging to or part of the real nature of sth./sb. 固有的；内在的；本身的
16. mitigation	/ˌmɪtɪ'geɪʃn/	n.	(formal) a reduction in the unpleasantness, seriousness, or painfulness of something 减轻，缓解，缓和
17. notification	/ˌnəʊtɪfɪ'keɪʃ(ə)n/	n.	(formal) the act of giving or receiving official information about sth.通知；通

告；告示

18. provision	/prəˈvɪʒ(ə)n/	n.	a condition or an arrangement in a legal document（文件的）规定，条款
19. ratification	/ˌrætɪfɪˈkeɪʃ(ə)n/	n.	the act or process of making an agreement officially valid by voting for or signing it 正式批准
20. resilience	/rɪˈzɪliəns/	n.	the ability of people or things to feel better quickly after sth. unpleasant, such as shock, injury, etc. 快速恢复的能力；适应力；抗御力
21. robust	/rəʊˈbʌst/	adj.	strong and not likely to fail or become weak 稳健的，强劲的；富有活力的
22. stocktake	/ˈstɒkteɪk/	n.	a reassessment of one's current situation, progress, prospects, etc. 总结，反思；估计；盘点
23. transparency	/trænsˈpærənsi/	n.	the quality of sth., such as glass, that allows you to see through it 透明；透明性

24. as appropriate   酌情，视情况而定
25. global peaking   全球峰值
26. in conjunction with   与…协力
27. mutatis mutandis   加以必要变更；准用

Exercises

Exercise 1   Reading Comprehension

Directions: *Read Article 4 of **Paris Agreement**, and decide whether the following statements are true or false. Write T for true or F for false in the brackets in front of each statement.*

1. (   ) The best available science has been used to rapidly reduce greenhouse gas emissions in connection with sustainable development and poverty eradication.

2. (   ) Each Party will gradually increase its nationally determined contribution, depending on the increase of its national population, and reflect its greatest extent possible, reflecting its common but differentiated responsibilities and respective capacities.

3. (   ) Underdeveloped country Parties should continue to take the lead to strengthen their mitigation efforts to gradually achieve economy-wide emission reduction or limitation targets in accordance with different national circumstances.

4. (   ) Strategies, plans and actions for low greenhouse gas emissions development in

the least developed countries and big island developing States could be developed and communicated to reflect their circumstances.

5. (   ) As required, each Party should communicate a nationally determined contribution every four years.

6. (   ) Double counting is to be avoided when each Party accounts for anthropogenic emissions and removals corresponding to its nationally owned contributions, and it should promote environmental integrity, transparency, precision, completeness, comparability and consistency.

Exercise 2   Skimming and Scanning

Directions: *Read the following passage excerpted from* **Paris Agreement**. *At the end of the passage, there are six statements. Each statement contains information given in one of the paragraphs of the passage. Identify the paragraph from which the information is derived. Each paragraph is marked with a letter. You may choose a paragraph more than once. Answer the questions by writing the corresponding letter in the brackets in front of each statement.*

A) Developed country Parties shall provide financial resources to assist developing country Parties with respect to both mitigation and adaptation in continuation of their existing obligations under the *Convention*.

B) Other Parties are encouraged to provide or continue to provide such support voluntarily.

C) As part of a global effort, developed country Parties should continue to take the lead in mobilizing climate finance from a wide variety of sources, instruments and channels, noting the significant role of public funds, through a variety of actions, including supporting country-driven strategies, and taking into account the needs and priorities of developing country Parties. Such mobilization of climate finance should represent a progression beyond previous efforts.

D) The provision of scaled-up financial resources should aim to achieve a balance between adaptation and mitigation, taking into account country-driven strategies, and the priorities and needs of developing country Parties, especially those that are particularly vulnerable to the adverse effects of climate change and have significant capacity constraints, such as the least developed countries and small island developing States, considering the need for public and grant-based resources for adaptation.

E) Developed country Parties shall biennially communicate indicative quantitative and qualitative information related to paragraphs 1 and 3 of this Article, as applicable, including, as available, projected levels of public financial resources to be provided to developing country Parties. Other Parties providing resources are encouraged to communicate biennially such information on a voluntary basis.

F) The global stocktake referred to in Article 14 shall take into account the relevant information provided by developed country Parties and/or Agreement bodies on efforts related to climate finance.

G) Developed country Parties shall provide transparent and consistent information on support for developing country Parties provided and mobilized through public interventions biennially in accordance with the modalities, procedures and guidelines to be adopted by the Conference of the Parties serving as the meeting of the Parties to the *Paris Agreement*, at its first session, as stipulated in Article 13, paragraph 13. Other Parties are encouraged to do so.

H) The Financial Mechanism of the *Convention*, including its operating entities, shall serve as the financial mechanism of this Agreement.

I) The institutions serving this Agreement, including the operating entities of the Financial Mechanism of the *Convention*, shall aim to ensure efficient access to financial resources through simplified approval procedures and enhanced readiness support for developing country Parties, in particular for the least developed countries and small island developing States, in the context of their national climate strategies and plans.

1. (　　) Achieving a balance between adaptation and mitigation should be the goal of the provision of larger financial resources.

2. (　　) Certain indicative quantitative and qualitative information should be communicated every two years.

3. (　　) One aim of the bodies serving this Agreement is to guarantee that developing country Parties can obtain financial support for their national climate strategies and plans.

4. (　　) Developed country Parties should provide funding to help developing country Parties in both mitigation and adaptation.

5. (　　) The mobilization of climate finance should outperform previous efforts with a consideration of the needs and priorities of developing country Parties.

6. (　　) Besides developed country Parties, other Parties are also encouraged to provide obvious and consistent information every two years.

Exercise 3　Word Formation

Directions: *In this section, there are ten sentences from* **Paris Agreement**. *You are required to complete these sentences with the proper form of the words given in blanks.*

1. This Agreement will be implemented to reflect equity and the principle of common but differentiated responsibilities and respective _____, in the light of different national circumstances. (capable)

2. Assist developing country Parties in identifying effective adaptation practices, adaptation needs, _____, support provided and received for adaptation actions and efforts, and challenges and gaps, in a manner consistent with encouraging good practices. (prior)

3. The _____ of climate change impacts and vulnerability, with a view to formulating nationally determined prioritized actions, taking into account vulnerable people, places and ecosystems. (assess)

4. Each Party should, as appropriate, submit and update _____ an adaptation communication, which may include its priorities, implementation and support needs, plans and actions, without creating any additional burden for developing country Parties. (period)

5. Parties should enhance understanding, action and support, including through the

Warsaw International Mechanism, as appropriate, on a _____ and facilitative basis with respect to loss and damage associated with the adverse effects of climate change. (cooperate)

6. All Parties enhancing the capacity of developing country Parties to implement this Agreement, including through regional, bilateral and multilateral approaches, shall _____ communicate on these actions or measures on capacity-building. (regular)

7. Capacity-building activities shall be enhanced through appropriate institutional arrangements to support the _____ of this Agreement, including the appropriate institutional arrangements established under the *Convention* that serve this Agreement. (implement)

8. Mitigation co-benefits resulting from Parties' adaptation actions and/or _____ diversification plans can contribute to mitigation outcomes under this Article. (economy)

9. Strengthening scientific knowledge on climate, including research, _____ observation of the climate system and early warning systems, in a manner that informs climate services and supports decision-making. (system)

10. At any time after three years from the date on which this Agreement has entered into force for a Party, that Party may withdraw from this Agreement by giving written _____ to the Depositary. (notify)

Exercise 4   Translation

Section A

Directions: *Read Paris Agreement, and complete the sentences by translating into English the Chinese given in blanks.*

1. Increase the ability to adapt to the adverse impacts of climate change and foster climate resilience and _____ (促进温室气体低排放发展), in a manner that does not threaten food production.

2. _____ (使用国际转让的减缓成果) to achieve nationally determined contributions under this Agreement shall be voluntary and authorized by participating Parties.

3. _____ (加强、鼓励和扶持创新) is critical for an effective, long-term global response to climate change and promoting economic growth and sustainable development.

4. Parties recognize the importance of _____ (综合、整体和平衡的非市场方法) being available to Parties to assist in the implementation of their nationally determined contributions, in the context of sustainable development and poverty eradication, in a coordinated and effective manner.

5. Capacity-building should be country-driven, based on and _____ _____ (响应国家的需要), and foster country ownership of Parties, in particular, for developing country Parties, including at the national, subnational and local levels.

6. Recognize the fundamental priority of _____ (保障粮食安全和消除饥饿), and the particular vulnerabilities of food production systems to the adverse impacts of climate change.

7. The Conference of the Parties serving as the meeting of the Parties to the *Paris*

Agreement may ＿＿＿＿＿＿＿＿＿＿＿＿＿＿＿＿ (为这些附属机构和体制安排提供进一步指导).

8. Parties hereby establish the global goal on adaptation of enhancing adaptive capacity, strengthening resilience and ＿＿＿＿＿＿＿＿＿＿＿＿＿＿＿ (减少对气候变化的脆弱性), with a view to contributing to sustainable development and ensuring an adequate adaptation response in the context of the temperature goal referred to in Article 2.

9. Parties recognize the importance of averting, minimizing and addressing loss and damage associated with the adverse effects of climate change, including ＿＿＿＿＿＿＿＿＿＿＿＿＿＿＿＿＿ (极端气候事件和缓发事件), and the role of sustainable development in reducing the risk of loss and damage.

10. In order to ＿＿＿＿＿＿＿＿＿＿＿＿＿＿＿＿ (建立互信并促进有效执行), an enhanced transparency framework for action and support, with built-in flexibility which takes into account Parties' different capacities and builds upon collective experience is hereby established.

Section B

Directions: *Translate the following sentences from English into Chinese.*

1. Hold the increase in the global average temperature to well below 2℃ above pre-industrial levels and to pursue efforts to limit the temperature increase to 1.5℃ above pre-industrial levels, recognizing that this would significantly reduce the risks and impacts of climate change. (*Article 2*)

2. Parties recognize that some Parties choose to pursue voluntary cooperation in the implementation of their nationally determined contributions to allow for higher ambition in their mitigation and adaptation actions and to promote sustainable development and environmental integrity. (*Article 6*)

3. Support, including financial support, shall be provided to developing country Parties for the implementation of this Article, including for strengthening cooperative action on technology development and transfer at different stages of the technology cycle, with a view to achieving a balance between support for mitigation and adaptation. (*Article 6*)

4. Parties, noting the importance of technology for the implementation of mitigation and adaptation actions under this Agreement and recognizing existing technology deployment and dissemination efforts, shall strengthen cooperative action on technology development and transfer. (*Article 10*)

5. In the case of regional economic integration organizations with one or more member States that are Parties to this Agreement, the organization and its member States shall decide on their respective responsibilities for the performance of their obligations under this Agreement. (*Article 20*)

6. Parties shall cooperate in taking measures, as appropriate, to enhance climate change education, training, public awareness, public participation and public access to information, recognizing the importance of these steps with respect to enhancing actions under this Agreement. (*Article 12*)

Extensive Readings

Passage 1

Directions: *Read the following passage and choose the best answer for each of the following questions according to the information given in the passage.*

As there is a direct relation between global average temperatures and the concentration of greenhouse gases (GHGs) in the atmosphere, the key for the solution to the climate change problem rests in decreasing the amount of emissions released into the atmosphere and in reducing the current concentration of carbon dioxide (CO_2) by enhancing sinks (e.g. increasing the area of forests). Efforts to reduce emissions and enhance sinks are referred to as "mitigation".

The *United Nations Framework Convention on Climate Change (*UNFCCC or *Convention*) requires all Parties, keeping in mind their responsibilities and capabilities, to formulate and implement programs containing measures to mitigate climate change. Such programs target economic activity with an aim to incentivize actions that are cleaner or disincentive (抑制) those that result in large amounts of GHGs. They include policies, incentives schemes and investment programs which address all sectors, including energy generation and use, transport, buildings, industry, agriculture, forestry and other land use, and waste management. Mitigation measures are translated in, for example, an increased use of renewable energy, the application of new technologies such as electric cars, or changes in practices or behaviors, such as driving less or changing one's diet. Further, they include expanding forests and other sinks to remove greater amounts of CO_2 from the atmosphere, or simply making improvements to a cook stove design.

Under the UNFCCC, and notably under the *Kyoto Protocol*, developed countries have set economy-wide caps for their national emissions, while developing countries have generally focused on specific programs and projects. Following the 2009 *Copenhagen Accord* and the 2010 *Cancun Agreements* developed countries have communicated quantified economy-wide emission targets for 2020 and developing countries have agreed to implement nationally appropriate mitigation actions with support from developed countries. In addition, developed country Parties to the *Kyoto Protocol* – at the end of the first commitment period under the Protocol (2008—2012) – adopted a second commitment period with targets for 2013—2020, in the form of the *Doha Amendment*. For developing countries the *Kyoto Protocol*'s clean development mechanism has been an important avenue of action for these countries to implement project activities that reduce emissions and enhance sinks.

In the process leading up to the Paris Conference all countries, developed and developing, prepared intended nationally determined contributions (INDCs), which outline national efforts to reduce emissions and increase resilience. As a result, a diversity of efforts was communicated, including absolute and relative quantified national targets, sectoral targets and programs, and others. The new concept of INDCs was eventually formalized under the *Paris Agreement* as nationally determined contributions (NDCs), and Parties are requested to prepare and communicate successive NDCs every five years.

Parties to the *Convention* have also cooperated increasingly to reduce GHG emissions from deforestation in developing countries. Developing countries are encouraged to contribute to mitigation actions in the forest sector by undertaking activities to reduce emissions from deforestation and forest degradation, conserve forest carbon stocks, implement sustainable management of forests and enhance forest carbon stocks(REDD-plus). The *Paris Agreement* also recognizes the importance of sinks, including forests and encourages Parties to implement and support the existing framework of guidance and decisions that has been elaborated on REDD-plus under the *Convention* over the years.

Emissions from international aviation and maritime transport contribute increasingly to global emissions. To address these emissions, there has been ongoing work in the International Civil Aviation Organization and the International Maritime Organization, as well as cooperation between these two organizations and the UNFCCC.

All over the world, many measures are being taken to mitigate climate change by countries trying to live up to their commitments under the *Convention*, the *Kyoto Protocol* and the *Paris Agreement*. According to the *Convention*, Parties shall take into consideration the specific needs and concerns of developing country Parties arising from the impacts of response measures, a call that is echoed similarly by the *Paris Agreement*. The *Kyoto Protocol* commits Parties to strive to minimize adverse economic, social and environmental impacts on other Parties, especially developing country Parties. In order to facilitate assessment and analysis such impacts, and with the view to recommending specific actions, the COP has established a forum on the impact of the implementation of response measures under the *Convention*, which is also to serve the *Paris Agreement*.

Climate change mitigation has been a central element in the intergovernmental negotiations carried out under the UNFCCC process. Negotiations on various items dealing with different aspects of mitigation pursuant to the *Convention*, the *Kyoto Protocol* and the *Paris Agreement* are currently ongoing under the negotiating bodies under the UNFCCC. Take *Paris Agreement* as an example. Mitigation lies at the heart of Parties' efforts to achieve the overall purpose and long-term temperature goals set out in Article 2 of the *Paris Agreement* of holding the increase in the global average temperature to well below 2℃ above pre-industrial levels and pursuing efforts to limit the temperature increase to 1.5℃. Under the *Paris Agreement*, each Party is required to put forward successive and progressively more ambitious nationally determined contributions (NDCs), representing its highest possible mitigation ambition. As part of the *Paris Agreement* Work Program Parties are currently considering further guidance for NDCs in relation to the mitigation section of decision 1/CP.21 under the APA, common time frames for NDCs referred to in Article 4, paragraph 10, of the *Paris Agreement* under the Subsidiary Body for Implementation (SBI), matters relating to Article 6 of the *Paris Agreement* under the Subsidiary Body for Scientific and Technological

Advice (SBSTA) and modalities, work program and functions of the forum on the impact of the implementation of response measures under the SBI and the SBSTA.

Under the *Paris Agreement* and in accordance with Article 4, paragraph 19, Parties should strive to formulate and communicate long-term low greenhouse gas emission development strategies (LT-LEDS) and are invited according to decision 1/CP 21 to communicate these, by 2020, to the secretariat.

The 2050 Pathways Platform, LEDS Global Partnership, the NDC Partnership, the UN Development Program, and the World Resources Institute, in cooperation with the UNFCCC jointly organized a workshop on 10—12 July 2018 in Bangkok to help Parties kick off the discussion on formulating the LT-LEDS. The objective of the workshop was to: ①Engage country governments now in the development of Long-term Strategies by outlining importance, advantages and urgency of the strategies. ②Highlight emerging good practice, country experiences, expert views and support for developing Long-term Strategies, as well as related challenges and how to overcome them. ③Explore the relationship between long-term impacts and near-term climate actions and NDC implementation. ④Build a community of practitioners and support for advancing the development of Long-term Strategies.

1. What is the key to approaching the climate change problem?＿＿＿＿
    A. Decreasing the number of heavy-industry enterprises and the number of transportation vehicles.
    B. Increasing the public's awareness of the hazard of climate change problem.
    C. Implementing sustainable management of forests and expanding forests.
    D. Cutting down emissions into the atmosphere and reducing the present concentration of CO_2.

2. According to the third paragraph, which of the following instruments proposed the clean development mechanism that has become a significant means of action for developing countries to implement projects to decrease emissions and enhance sinks?＿＿＿＿
    A. The *United Nations Framework Convention on Climate Change*
    B. *Kyoto Protocol*
    C. *Copenhagen Accord*
    D. *Paris Agreement*

3. For what purpose are the International Civil Aviation Organization and the International Maritime Organization mentioned in this passage?＿＿＿＿
    A. To show that they are two alternatives to the land transportation vehicles which may cause great emissions.
    B. To illustrate that the two organizations have made great contributions to decrease the increasing global emissions.
    C. To indicate that they should also take actions since international aviation and maritime transport also cause increasing global emissions.
    D. To prove that it is mainly the responsibility of land organizations rather than aviation and maritime organizations to address global emissions.

4. According to the eighth paragraph, what is the core task for the achievement of the overall purpose and long-term temperature goals set out in *Paris Agreement*?＿＿＿＿
   A. Mitigation.
   B. Energy generation.
   C. Sustainable forest management.
   D. Long-term investment of all parties.

5. According to the last two paragraphs, to which institution should all parties of *Paris Agreement* communicate their long-term low greenhouse gas emission development strategies? ＿＿＿＿
   A. The World Resources Institute
   B. The UN Development Program
   C. The secretariat (of the 2015 Paris Conference on Climate Change)
   D. World Green Climate Association

Passage 2

Directions: *In this section, there is a passage with twelve blanks. You are required to select one word for each blank from a list of choices given in a word bank following the passage. Read the passage through carefully before making your choices. Each choice in the bank is identified by a letter. You may not use any of the words in the bank more than once.*

A. objectives	B. initiatives	C. maintaining	D. released
E. imperative	F. biodiversity	G. addressing	H. solution
I. sustainable	J. degradation	K. globally	L. subnational

Forests are a stabilizing force for the climate. They regulate ecosystems, protect __1__, play an integral part in the carbon cycle, support livelihoods, and supply goods and services that can drive sustainable growth. Forests' role in climate change is two-fold. They act as both a cause and a __2__ for greenhouse gas emissions. Around 25% of global emissions come from the land sector, the second largest source of greenhouse gas emissions after the energy sector. About half of these comes from deforestation and forest __3__.

Forests are also one of the most important solutions to __4__ the effects of climate change. Approximately 2.6 billion tons of carbon dioxide, one-third of the CO_2 __5__ from burning fossil fuels, is absorbed by forests every year. Estimates show that nearly two billion hectares of degraded land across the world — an area the size of South America—offer opportunities for restoration. Increasing and __6__ forests is therefore an essential solution to climate change.

Halting the loss and degradation of forest ecosystems and promoting their restoration have the potential to contribute over one-third of the total climate change mitigation that scientists say is required by 2030 to meet the __7__ of the *Paris Agreement*.

Today, more and more consumers are demanding forest products from __8__ sources, and an increasing number of major palm oil, timber, paper and other forest product corporations are beginning the conversion to deforestation-free supply chains.

In addition to creating and maintaining protected areas and launching __9__ towards more sustainable management, many countries, __10__ governments and private landowners are restoring degraded and deforested land. This helps to take pressure off healthy, intact forests and reduce emissions from deforestation and forest degradation.

As the world debates how to operationalise the *Paris Agreement*, it is __11__ that national leaders accelerate these actions. This can be done by subscribing to and implementing the *New York Declaration on Forests*, sustain forest climate financing, and include forest and land use in countries' Nationally Determined Contributions under the *Paris Agreement*.

Nature—and in particular, trees and forests—can and must be part of the solution to keeping the climate within the __12__ accepted two-degree temperature increase limit.

Further Studies and Post-Reading Discussion

Task 1
Directions: *Surf the Internet and find more information about* **Paris Agreement**. *Work in groups and work out a report on one of the following topics.*

1. The *Paris Agreement*'s central aim.
2. China's efforts in climate changes actions.

Task 2
Directions: *Read the following sentences on Eco-Civilization and make a speech on your understanding of the eco-environmental conservation.*

山水林田湖草综合治理

山水林田湖草是一个生命共同体。人的命脉在田，田的命脉在水，水的命脉在山，山的命脉在土，土的命脉在林和草，这个生命共同体是人类生存发展的物质基础。统筹山水林田湖草系统治理，就是坚持山水林田湖草是生命共同体，按照生态系统的整体性、系统性及其内在规律，统筹考虑自然生态各要素、山上山下、地上地下、陆地海洋以及流域上下游，进行整体保护、系统修复、综合治理，增强生态系统循环能力，维护生态平衡。（摘自《中国关键词》生态文明篇）	Mountains, rivers, forests, farmlands, lakes and grasslands are a community of life. The lifeline of humans rests with farmlands; that of farmlands with water; that of water with mountains; that of mountains with earth; and that of earth with forests and grasslands. This community of life is the material foundation for the survival and development of humanity. This holistic approach follows the innate laws of the ecosystems to achieve systematic protection and restoration, giving consideration to a full range of environmental factors -up and down the mountains, above and under the ground, on the land and in the oceans, in the upper and lower reaches of the river basins, so as to increase the circulation of ecosystems and maintain ecological balance. (Excerpt from *Keywords to Understand China on Eco-Civilization*)

第 6 章 改变我们的世界：2030 年可持续发展议程

Chapter 6 Transforming Our World: the 2030 Agenda for Sustainable Development

Background and Significance

2015 年 9 月 25 日至 27 日，联合国 193 个成员国在可持续发展峰会上正式通过成果性文件——《改变我们的世界：2030 年可持续发展议程》（简称《2030 年可持续发展议程》）。这一涵盖 17 项可持续发展目标（简称 SDGs）和 169 项具体目标的纲领性文件旨在推动未来 15 年内实现 3 项宏伟的全球目标：消除极端贫困、战胜不平等和不公正、遏制气候变化及保护环境。

《2030 年可持续发展议程》源于千年发展目标（Millennium Development Goals, MDGs）。2000 年 9 月，189 个国家的代表在联合国千年峰会上通过千年发展目标，所有目标以 1990 年为基础，完成时间为 2015 年。千年发展目标包括消除贫穷、饥饿、疾病、文盲、环境恶化和对妇女的歧视等 8 项目标。据联合国发布的最后一份千年发展目标差距问题工作组报告显示，尽管千年发展目标已经取得了很大的进展，但是目标框架中很多目标仍然没有实现，贫困仍然是主要挑战和优先事项，两性平等、环境可持续发展和全球伙伴关系等目标仍然进展缓慢，而这一局面与千年发展目标自身的缺陷密切相关。所有的研究都指出一个事实：千年发展目标的设计框架没有在经济、社会和环境 3 个相互联系的领域充分处理可持续发展问题。在 2015 年千年发展目标收官之际，联合国全体成员通过了《2030 年可持续发展议程》，希望到 2030 年实现经济增长、社会包容与环境的可持续性三者的和谐发展，体现了国际社会团结一致，应对人类社会发展重大挑战的决心。

《2030 年可持续发展议程》以可持续发展目标为核心议程，在千年发展目标的基础构架上，从设计上超越传统可持续发展的经济、社会、环境三维认知，增加了公正保障和执行手段，上升至"5P"即人(People)、地球(Planet)、繁荣(Prosperity)、和平(Peace)和伙伴关系(Partnership)，从发展理念上创新丰富了可持续发展的维度。

Transforming Our World: the 2030 Agenda for Sustainable Development

Preamble

This Agenda is a plan of action for people, planet and prosperity. It also seeks to strengthen universal peace in larger freedom. We recognise that eradicating poverty in all its forms and dimensions, including extreme poverty, is the greatest global challenge and an indispensable requirement for sustainable development.

All countries and all stakeholders, acting in collaborative partnership, will implement this plan. We are resolved to free the human race from the tyranny of poverty and want and to heal and secure our planet. We are determined to take the bold and transformative steps which are urgently needed to shift the world onto a sustainable and resilient path. As we embark on this collective journey, we pledge that no one will be left behind.

The 17 Sustainable Development Goals and 169 targets which we are announcing today demonstrate the scale and ambition of this new universal Agenda. They seek to build on the Millennium Development Goals and complete what these did not achieve. They seek to realize the human rights of all and to achieve gender equality and the empowerment of all women and girls. They are integrated and indivisible and balance the three dimensions of sustainable development: the economic, social and environmental.

The Goals and targets will stimulate action over the next fifteen years in areas of critical importance for humanity and the planet.

People

We are determined to end poverty and hunger, in all their forms and dimensions, and to ensure that all human beings can fulfil their potential in dignity and equality and in a healthy environment.

Planet

We are determined to protect the planet from degradation, including through sustainable consumption and production, sustainably managing its natural resources and taking urgent action on climate change, so that it can support the needs of the present and future generations.

Prosperity

We are determined to ensure that all human beings can enjoy prosperous and fulfilling lives and that economic, social and technological progress occurs in harmony with nature.

Peace

We are determined to foster peaceful, just and inclusive societies which are free from fear and violence. There can be no sustainable development without peace and no peace without sustainable development.

Partnership

We are determined to mobilize the means required to implement this Agenda through a revitalised Global Partnership for Sustainable Development, based on a spirit of strengthened global solidarity, focussed in particular on the needs of the poorest and most vulnerable and with the participation of all countries, all stakeholders and all people.

The interlinkages and integrated nature of the Sustainable Development Goals are of crucial importance in ensuring that the purpose of the new Agenda is realised. If we realize our ambitions across the full extent of the Agenda, the lives of all will be profoundly improved and our world will be transformed for the better.

DECLARATION

Introduction

1. We, the Heads of State and Government and High Representatives, meeting at the United Nations Headquarters in New York from 25—27 September 2015 as the Organization celebrates its seventieth anniversary, have decided today on new global Sustainable Development Goals.

2. On behalf of the peoples we serve, we have adopted a historic decision on a comprehensive, far-reaching and people-centred set of universal and transformative Goals and targets. We commit ourselves to working tirelessly for the full implementation of this Agenda by 2030. We recognize that eradicating poverty in all its forms and dimensions, including extreme poverty, is the greatest global challenge and an indispensable requirement for sustainable development. We are committed to achieving sustainable development in its three dimensions—economic, social and environmental—in a balanced and integrated manner. We will also build upon the achievements of the Millennium Development Goals and seek to address their unfinished business.

3. We resolve, between now and 2030, to end poverty and hunger everywhere; to combat inequalities within and among countries; to build peaceful, just and inclusive societies; to protect human rights and promote gender equality and the empowerment of women and girls; and to ensure the lasting protection of the planet and its natural resources. We resolve also to create conditions for sustainable, inclusive and sustained economic growth, shared prosperity and decent work for all, taking into account different levels of national development and capacities.

4. As we embark on this great collective journey, we pledge that no one will be left behind. Recognizing that the dignity of the human person is fundamental, we wish to see the Goals and targets met for all nations and peoples and for all segments of society. And we will endeavour to reach the furthest behind first.

5. This is an Agenda of unprecedented scope and significance. It is accepted by all countries and is applicable to all, taking into account different national realities, capacities and levels of development and respecting national policies and priorities. These are universal goals and targets which involve the entire world, developed and developing countries alike. They

are integrated and indivisible and balance the three dimensions of sustainable development.

6. The Goals and targets are the result of over two years of intensive public consultation and engagement with civil society and other stakeholders around the world, which paid particular attention to the voices of the poorest and most vulnerable. This consultation included valuable work done by the General Assembly Open Working Group on Sustainable Development Goals and by the United Nations, whose Secretary-General provided a synthesis report in December 2014.

Our vision

7. In these Goals and targets, we are setting out a supremely ambitious and transformational vision. We envisage a world free of poverty, hunger, disease and want, where all life can thrive. We envisage a world free of fear and violence. A world with universal literacy. A world with equitable and universal access to quality education at all levels, to health care and social protection, where physical, mental and social well-being are assured. A world where we reaffirm our commitments regarding the human right to safe drinking water and sanitation and where there is improved hygiene; and where food is sufficient, safe, affordable and nutritious. A world where human habitats are safe, resilient and sustainable and where there is universal access to affordable, reliable and sustainable energy.

8. We envisage a world of universal respect for human rights and human dignity, the rule of law, justice, equality and non-discrimination; of respect for race, ethnicity and cultural diversity; and of equal opportunity permitting the full realization of human potential and contributing to shared prosperity. A world which invests in its children and in which every child grows up free from violence and exploitation. A world in which every woman and girl enjoys full gender equality and all legal, social and economic barriers to their empowerment have been removed. A just, equitable, tolerant, open and socially inclusive world in which the needs of the most vulnerable are met.

9. We envisage a world in which every country enjoys sustained, inclusive and sustainable economic growth and decent work for all. A world in which consumption and production patterns and use of all natural resources – from air to land, from rivers, lakes and aquifers to oceans and seas—are sustainable. One in which democracy, good governance and the rule of law as well as an enabling environment at national and international levels, are essential for sustainable development, including sustained and inclusive economic growth, social development, environmental protection and the eradication of poverty and hunger. One in which development and the application of technology are climate-sensitive, respect biodiversity and are resilient. One in which humanity lives in harmony with nature and in which wildlife and other living species are protected.

Our shared principles and commitments

10. The new Agenda is guided by the purposes and principles of the *Charter of the United Nations*, including full respect for international law. It is grounded in the *Universal Declaration of Human Rights*[①], international human rights treaties, the *Millennium*

---

① Resolution 217 A (III).

Declaration[①] and the 2005 *World Summit Outcome Document*[②]. It is informed by other instruments such as the *Declaration on the Right to Development*[③].

11. We reaffirm the outcomes of all major UN conferences and summits which have laid a solid foundation for sustainable development and have helped to shape the new Agenda. These include the *Rio Declaration on Environment and Development*[④]; the World Summit on Sustainable Development; the World Summit for Social Development; *the Programme of Action of the International Conference on Population and Development*[⑤], the Beijing Platform for Action[⑥]; and the United Nations Conference on Sustainable Development ("Rio+ 20"). We also reaffirm the follow-up to these conferences, including the outcomes of the Fourth United Nations Conference on the Least Developed Countries, the Third International Conference on Small Island Developing States; the Second United Nations Conference on Landlocked Developing Countries; and the Third UN World Conference on Disaster Risk Reduction.

12. We reaffirm all the principles of the *Rio Declaration on Environment and Development*, including, inter alia, the principle of common but differentiated responsibilities, as set out in principle 7 thereof.

13. The challenges and commitments contained in these major conferences and summits are interrelated and call for integrated solutions. To address them effectively, a new approach is needed. Sustainable development recognizes that eradicating poverty in all its forms and dimensions, combatting inequality within and among countries, preserving the planet, creating sustained, inclusive and sustainable economic growth and fostering social inclusion are linked to each other and are interdependent.

Our world today

14. We are meeting at a time of immense challenges to sustainable development. Billions of our citizens continue to live in poverty and are denied a life of dignity. There are rising inequalities within and among countries. There are enormous disparities of opportunity, wealth and power. Gender inequality remains a key challenge. Unemployment, particularly youth unemployment, is a major concern. Global health threats, more frequent and intense natural disasters, spiralling conflict, violent extremism, terrorism and related humanitarian crises and forced displacement of people threaten to reverse much of the development

---

① Resolution 55/2.

② Resolution 60/1.

③ Resolution 41/128, annex.

④ Report of the United Nations Conference on Environment and Development, Rio de Janeiro, 3–14 June 1992, vol. I, Resolutions Adopted by the Conference (United Nations publication, Sales No. E.93.I.8 and corrigendum), resolution 1, annex I.

⑤ Report of the International Conference on Population and Development, Cairo, 5–13 September 1994 (United Nations publication, Sales No. E.95.XIII.18), chap. I, resolution 1, annex.

⑥ Report of the Fourth World Conference on Women, Beijing, 4–15 September 1995 (United Nations publication, Sales No. E.96.IV.13), chap. I, resolution 1, annex II.

progress made in recent decades. Natural resource depletion and adverse impacts of environmental degradation, including desertification, drought, land degradation, freshwater scarcity and loss of biodiversity, add to and exacerbate the list of challenges which humanity faces. Climate change is one of the greatest challenges of our time and its adverse impacts undermine the ability of all countries to achieve sustainable development. Increases in global temperature, sea level rise, ocean acidification and other climate change impacts are seriously affecting coastal areas and low-lying coastal countries, including many least developed countries and small island developing States. The survival of many societies, and of the biological support systems of the planet, is at risk.

15. It is also, however, a time of immense opportunity. Significant progress has been made in meeting many development challenges. Within the past generation, hundreds of millions of people have emerged from extreme poverty. Access to education has greatly increased for both boys and girls. The spread of information and communications technology and global interconnectedness has great potential to accelerate human progress, to bridge the digital divide and to develop knowledge societies, as does scientific and technological innovation across areas as diverse as medicine and energy.

16. Almost fifteen years ago, the Millennium Development Goals were agreed. These provided an important framework for development and significant progress has been made in a number of areas. But the progress has been uneven, particularly in Africa, least developed countries, landlocked developing countries, and small island developing States, and some of the MDGs remain off-track, in particular those related to maternal, newborn and child health and to reproductive health. We recommit ourselves to the full realization of all the MDGs, including the off-track MDGs, in particular by providing focussed and scaled-up assistance to least developed countries and other countries in special situations, in line with relevant support programmes. The new Agenda builds on the Millennium Development Goals and seeks to complete what these did not achieve, particularly in reaching the most vulnerable.

17. In its scope, however, the framework we are announcing today goes far beyond the MDGs. Alongside continuing development priorities such as poverty eradication, health, education and food security and nutrition, it sets out a wide range of economic, social and environmental objectives. It also promises more peaceful and inclusive societies. It also, crucially, defines means of implementation. Reflecting the integrated approach that we have decided on, there are deep interconnections and many cross-cutting elements across the new Goals and targets.

The new Agenda

18. We are announcing today 17 Sustainable Development Goals with 169 associated targets which are integrated and indivisible. Never before have world leaders pledged common action and endeavour across such a broad and universal policy agenda. We are setting out together on the path towards sustainable development, devoting ourselves collectively to the pursuit of global development and of "win-win" cooperation which can bring huge gains to all countries and all parts of the world. We reaffirm that every State has,

and shall freely exercise, full permanent sovereignty over all its wealth, natural resources and economic activity. We will implement the Agenda for the full benefit of all, for today's generation and for future generations. In doing so, we reaffirm our commitment to international law and emphasize that the Agenda is to be implemented in a manner that is consistent with the rights and obligations of states under international law.

19. We reaffirm the importance of the *Universal Declaration of Human Rights*, as well as other international instruments relating to human rights and international law. We emphasize the responsibilities of all States, in conformity with the *Charter of the United Nations*, to respect, protect and promote human rights and fundamental freedoms for all, without distinction of any kind as to race, colour, sex, language, religion, political or other opinion, national or social origin, property, birth, disability or other status.

20. Realizing gender equality and the empowerment of women and girls will make a crucial contribution to progress across all the Goals and targets. The achievement of full human potential and of sustainable development is not possible if one half of humanity continues to be denied its full human rights and opportunities. Women and girls must enjoy equal access to quality education, economic resources and political participation as well as equal opportunities with men and boys for employment, leadership and decision-making at all levels. We will work for a significant increase in investments to close the gender gap and strengthen support for institutions in relation to gender equality and the empowerment of women at the global, regional and national levels. All forms of discrimination and violence against women and girls will be eliminated, including through the engagement of men and boys. The systematic mainstreaming of a gender perspective in the implementation of the Agenda is crucial.

21. The new Goals and targets will come into effect on 1 January 2016 and will guide the decisions we take over the next fifteen years. All of us will work to implement the Agenda within our own countries and at the regional and global levels, taking into account different national realities, capacities and levels of development and respecting national policies and priorities. We will respect national policy space for sustained, inclusive and sustainable economic growth, in particular for developing states, while remaining consistent with relevant international rules and commitments. We acknowledge also the importance of the regional and sub-regional dimensions, regional economic integration and interconnectivity in sustainable development. Regional and sub-regional frameworks can facilitate the effective translation of sustainable development policies into concrete action at national level.

22. Each country faces specific challenges in its pursuit of sustainable development. The most vulnerable countries and, in particular, African countries, least developed countries, landlocked developing countries and small island developing states deserve special attention, as do countries in situations of conflict and post-conflict countries. There are also serious challenges within many middle-income countries.

23. People who are vulnerable must be empowered. Those whose needs are reflected in the Agenda include all children, youth, persons with disabilities (of whom more than 80% live

in poverty), people living with HIV/AIDS, older persons, indigenous peoples, refugees and internally displaced persons and migrants. We resolve to take further effective measures and actions, in conformity with international law, to remove obstacles and constraints, strengthen support and meet the special needs of people living in areas affected by complex humanitarian emergencies and in areas affected by terrorism.

24. We are committed to ending poverty in all its forms and dimensions, including by eradicating extreme poverty by 2030. All people must enjoy a basic standard of living, including through social protection systems. We are also determined to end hunger and to achieve food security as a matter of priority and to end all forms of malnutrition. In this regard, we reaffirm the important role and inclusive nature of the Committee on World Food Security and welcome the *Rome Declaration on Nutrition and Framework for Action*[①]. We will devote resources to developing rural areas and sustainable agriculture and fisheries, supporting smallholder farmers, especially women farmers, herders and fishers in developing countries, particularly least developed countries.

25. We commit to providing inclusive and equitable quality education at all levels—early childhood, primary, secondary, tertiary, technical and vocational training. All people, irrespective of sex, age, race, ethnicity, and persons with disabilities, migrants, indigenous peoples, children and youth, especially those in vulnerable situations, should have access to life-long learning opportunities that help them acquire the knowledge and skills needed to exploit opportunities and to participate fully in society. We will strive to provide children and youth with a nurturing environment for the full realization of their rights and capabilities, helping our countries to reap the demographic dividend including through safe schools and cohesive communities and families.

26. To promote physical and mental health and well-being, and to extend life expectancy for all, we must achieve universal health coverage and access to quality health care. No one must be left behind. We commit to accelerating the progress made to date in reducing newborn, child and maternal mortality by ending all such preventable deaths before 2030. We are committed to ensuring universal access to sexual and reproductive health-care services, including for family planning, information and education. We will equally accelerate the pace of progress made in fighting malaria, HIV/AIDS, tuberculosis, hepatitis, Ebola and other communicable diseases and epidemics, including by addressing growing anti-microbial resistance and the problem of unattended diseases affecting developing countries. We are committed to the prevention and treatment of non-communicable diseases, including behavioural, developmental and neurological disorders, which constitute a major challenge for sustainable development.

27. We will seek to build strong economic foundations for all our countries. Sustained, inclusive and sustainable economic growth is essential for prosperity. This will only be possible if wealth is shared and income inequality is addressed. We will work to build

---

① World Health Organization, document EB 136/8, annexes I and II.

dynamic, sustainable, innovative and people-centred economies, promoting youth employment and women's economic empowerment, in particular, and decent work for all. We will eradicate forced labour and human trafficking and end child labour in all its forms. All countries stand to benefit from having a healthy and well-educated workforce with the knowledge and skills needed for productive and fulfilling work and full participation in society. We will strengthen the productive capacities of least-developed countries in all sectors, including through structural transformation. We will adopt policies which increase productive capacities, productivity and productive employment; financial inclusion; sustainable agriculture, pastoralist and fisheries development; sustainable industrial development; universal access to affordable, reliable, sustainable and modern energy services; sustainable transport systems; and quality and resilient infrastructure.

28. We commit to making fundamental changes in the way that our societies produce and consume goods and services. Governments, international organizations, the business sector and other non-state actors and individuals must contribute to changing unsustainable consumption and production patterns, including through the mobilization, from all sources, of financial and technical assistance to strengthen developing countries' scientific, technological and innovative capacities to move towards more sustainable patterns of consumption and production. We encourage the implementation of the 10-Year Framework of Programmes on Sustainable Consumption and Production. All countries take action, with developed countries taking the lead, taking into account the development and capabilities of developing countries.

29. We recognize the positive contribution of migrants for inclusive growth and sustainable development. We also recognize that international migration is a multi-dimensional reality of major relevance for the development of countries of origin, transit and destination, which requires coherent and comprehensive responses. We will cooperate internationally to ensure safe, orderly and regular migration involving full respect for human rights and the humane treatment of migrants regardless of migration status, of refugees and of displaced persons. Such cooperation should also strengthen the resilience of communities hosting refugees, particularly in developing countries. We underline the right of migrants to return to their country of citizenship, and recall that States must ensure that their returning nationals are duly received.

30. States are strongly urged to refrain from promulgating and applying any unilateral economic, financial or trade measures not in accordance with international law and the *Charter of the United Nations* that impede the full achievement of economic and social development, particularly in developing countries.

31. We acknowledge that the UNFCCC[①] is the primary international, intergovernmental forum for negotiating the global response to climate change. We are determined to address decisively the threat posed by climate change and environmental degradation. The global nature of climate change calls for the widest possible international cooperation aimed at accelerating

---

① United Nations, Treaty Series, vol. 1771, No. 30822.

the reduction of global greenhouse gas emissions and addressing adaptation to the adverse impacts of climate change. We note with grave concern the significant gap between the aggregate effect of Parties' mitigation pledges in terms of global annual emissions of greenhouse gases by 2020 and aggregate emission pathways consistent with having a likely chance of holding the increase in global average temperature below 2℃ or 1.5℃ above pre-industrial levels.

32. Looking ahead to the twenty-first session of the Conference of the Parties, we underscore the commitment of all States to work for an ambitious and universal climate agreement. We reaffirm that the protocol, another legal instrument or agreed outcome with legal force under the *Convention* applicable to all Parties shall address in a balanced manner, inter alia, mitigation, adaptation, finance, technology development and transfer, and capacity-building, and transparency of action and support.

33. We recognise that social and economic development depends on the sustainable management of our planet's natural resources. We are therefore determined to conserve and sustainably use oceans and seas, freshwater resources, as well as forests, mountains and drylands and to protect biodiversity, ecosystems and wildlife. We are also determined to promote sustainable tourism, tackle water scarcity and water pollution, to strengthen cooperation on desertification, dust storms, land degradation and drought and to promote resilience and disaster risk reduction. In this regard, we look forward to the thirteenth meeting of the Parties to the *Convention on Biological Diversity* to be held in Mexico in 2016.

34. We recognize that sustainable urban development and management are crucial to the quality of life of our people. We will work with local authorities and communities to renew and plan our cities and human settlements so as to foster community cohesion and personal security and to stimulate innovation and employment. We will reduce the negative impacts of urban activities and of chemicals which are hazardous for human health and the environment, including through the environmentally sound management and safe use of chemicals, the reduction and recycling of waste and more efficient use of water and energy. And we will work to minimize the impact of cities on the global climate system. We will also take account of population trends and projections in our national, rural and urban development strategies and policies. We look forward to the upcoming United Nations Conference on Housing and Sustainable Urban Development in Quito, Ecuador.

35. Sustainable development cannot be realized without peace and security; and peace and security will be at risk without sustainable development. The new Agenda recognizes the need to build peaceful, just and inclusive societies that provide equal access to justice and that are based on respect for human rights (including the right to development), on effective rule of law and good governance at all levels and on transparent, effective and accountable institutions. Factors which give rise to violence, insecurity and injustice, such as inequality, corruption, poor governance and illicit financial and arms flows, are addressed in the Agenda. We must redouble our efforts to resolve or prevent conflict and to support post-conflict countries, including through ensuring that women have a role in peace-building and state-building. We call for further effective measures and actions to be taken, in conformity

with international law, to remove the obstacles to the full realization of the right of self-determination of peoples living under colonial and foreign occupation, which continue to adversely affect their economic and social development as well as their environment.

36. We pledge to foster inter-cultural understanding, tolerance, mutual respect and an ethic of global citizenship and shared responsibility. We acknowledge the natural and cultural diversity of the world and recognize that all cultures and civilizations can contribute to, and are crucial enablers of, sustainable development.

37. Sport is also an important enabler of sustainable development. We recognize the growing contribution of sport to the realization of development and peace in its promotion of tolerance and respect and the contributions it makes to the empowerment of women and of young people, individuals and communities as well as to health, education and social inclusion objectives.

38. We reaffirm, in accordance with the *Charter of the United Nations*, the need to respect the territorial integrity and political independence of States.

Means of Implementation

39. The scale and ambition of the new Agenda requires a revitalized Global Partnership to ensure its implementation. We fully commit to this. This Partnership will work in a spirit of global solidarity, in particular solidarity with the poorest and with people in vulnerable situations. It will facilitate an intensive global engagement in support of implementation of all the Goals and targets, bringing together Governments, the private sector, civil society, the United Nations system and other actors and mobilizing all available resources.

40. The means of implementation targets under Goal 17 and under each SDG are key to realising our Agenda and are of equal importance with the other Goals and targets. The Agenda, including the SDGs, can be met within the framework of a revitalized global partnership for sustainable development, supported by the concrete policies and actions as outlined in the outcome document of the Third International Conference on Financing for Development, held in Addis Ababa from 13—16 July 2015. We welcome the endorsement by the General Assembly of the *Addis Ababa Action Agenda*[①], which is an integral part of the *2030 Agenda for Sustainable Development*. We recognize that the full implementation of the *Addis Ababa Action Agenda* is critical for the realization of the Sustainable Development Goals and targets.

41. We recognize that each country has primary responsibility for its own economic and social development. The new Agenda deals with the means required for implementation of the Goals and targets. We recognize that these will include the mobilization of financial resources as well as capacity-building and the transfer of environmentally sound technologies to developing countries on favourable terms, including on concessional and preferential terms, as

---

① The *Addis Ababa Action Agenda* of the Third International Conference on Financing for Development (*Addis Ababa Action Agenda*), adopted by the General Assembly on 27 July 2015 (resolution 69/313, annex).

mutually agreed. Public finance, both domestic and international, will play a vital role in providing essential services and public goods and in catalyzing other sources of finance. We acknowledge the role of the diverse private sector, ranging from micro-enterprises to cooperatives to multinationals, and that of civil society organizations and philanthropic organizations in the implementation of the new Agenda.

42. We support the implementation of relevant strategies and programmes of action, including the Istanbul Declaration and Programme of Action[①], the *SIDS Accelerated Modalities of Action* (SAMOA) *Pathway*[②], the *Vienna Programme of Action for Landlocked Developing Countries for the Decade 2014—2024*[③], and reaffirm the importance of supporting the African Union's *Agenda 2063* and the programme of the New Partnership for Africa's Development (NEPAD)[④], all of which are integral to the new Agenda. We recognize the major challenge to the achievement of durable peace and sustainable development in countries in conflict and post-conflict situations.

43. We emphasize that international public finance plays an important role in complementing the efforts of countries to mobilize public resources domestically, especially in the poorest and most vulnerable countries with limited domestic resources. An important use of international public finance, including ODA, is to catalyse additional resource mobilization from other sources, public and private. ODA providers reaffirm their respective commitments, including the commitment by many developed countries to achieve the target of 0.7% of ODA/GNI to developing countries and 0.15% to 0.2% of ODA/GNI to least developed countries.

44. We acknowledge the importance for international financial institutions to support, in line with their mandates, the policy space of each country, in particular developing countries. We recommit to broadening and strengthening the voice and participation of developing countries—including African countries, least developed countries, land-locked developing countries, small-island developing States and middle-income countries—in international economic decision-making, norm-setting and global economic governance.

45. We acknowledge also the essential role of national parliaments through their enactment of legislation and adoption of budgets and their role in ensuring accountability for the effective implementation of our commitments. Governments and public institutions will also work closely on implementation with regional and local authorities, sub-regional institutions, international institutions, academia, philanthropic organisations, volunteer groups and others.

46. We underline the important role and comparative advantage of an adequately

---

① Report of the Fourth United Nations Conference on the Least Developed Countries, Istanbul, Turkey, 9–13 May 2011 (A/CONF.219/7), chaps. I and II.

② Resolution 69/15, annex.

③ Resolution 69/137, annex II.

④ A/57/304, annex.

resourced, relevant, coherent, efficient and effective UN system in supporting the achievement of the SDGs and sustainable development. While stressing the importance of strengthened national ownership and leadership at country level, we express our support for the ongoing ECOSOC Dialogue on the longer-term positioning of the United Nations development system in the context of this Agenda.

Follow-up and review

47. Our Governments have the primary responsibility for follow-up and review, at the national, regional and global levels, in relation to the progress made in implementing the Goals and targets over the coming fifteen years. To support accountability to our citizens, we will provide for systematic follow-up and review at the various levels, as set out in this Agenda and the *Addis Ababa Action Agenda*. The High Level Political Forum under the auspices of the General Assembly and the Economic and Social Council will have the central role in overseeing follow-up and review at the global level.

48. Indicators are being developed to assist this work. Quality, accessible, timely and reliable disaggregated data will be needed to help with the measurement of progress and to ensure that no one is left behind. Such data is key to decision-making. Data and information from existing reporting mechanisms should be used where possible. We agree to intensify our efforts to strengthen statistical capacities in developing countries, particularly African countries, least developed countries, landlocked developing countries, small island developing States and middle-income countries. We are committed to developing broader measures of progress to complement gross domestic product (GDP).

A call for action to change our world

49. Seventy years ago, an earlier generation of world leaders came together to create the United Nations. From the ashes of war and division they fashioned this Organization and the values of peace, dialogue and international cooperation which underpin it. The supreme embodiment of those values is the *Charter of the United Nations*.

50. Today we are also taking a decision of great historic significance. We resolve to build a better future for all people, including the millions who have been denied the chance to lead decent, dignified and rewarding lives and to achieve their full human potential. We can be the first generation to succeed in ending poverty; just as we may be the last to have a chance of saving the planet. The world will be a better place in 2030 if we succeed in our objectives.

51. What we are announcing today—an Agenda for global action for the next fifteen years—is a charter for people and planet in the 21st century. Children and young women and men are critical agents of change and will find in the new Goals a platform to channel their infinite capacities for activism into the creation of a better world.

52. " We the Peoples" are the celebrated opening words of the *UN Charter*. It is " We the Peoples" who are embarking today on the road to 2030. Our journey will involve Governments as well as Parliaments, the UN system and other international institutions, local authorities, indigenous peoples, civil society, business and the private sector, the scientific and academic community—and all people. Millions have already engaged with, and will own, this

Agenda. It is an Agenda of the people, by the people, and for the people—and this, we believe, will ensure its success.

53. The future of humanity and of our planet lies in our hands. It lies also in the hands of today's younger generation who will pass the torch to future generations. We have mapped the road to sustainable development; it will be for all of us to ensure that the journey is successful and its gains irreversible.

Sustainable Development Goals and targets

54. Following an inclusive process of intergovernmental negotiations, and based on the Proposal of the Open Working Group on Sustainable Development Goals[①], which includes a chapeau contextualising the latter, the following are the Goals and targets which we have agreed.

55. The SDGs and targets are integrated and indivisible, global in nature and universally applicable, taking into account different national realities, capacities and levels of development and respecting national policies and priorities. Targets are defined as aspirational and global, with each government setting its own national targets guided by the global level of ambition but taking into account national circumstances. Each government will also decide how these aspirational and global targets should be incorporated in national planning processes, policies and strategies. It is important to recognize the link between sustainable development and other relevant ongoing processes in the economic, social and environmental fields.

56. In deciding upon these Goals and targets, we recognise that each country faces specific challenges to achieve sustainable development, and we underscore the special challenges facing the most vulnerable countries and, in particular, African countries, least developed countries, landlocked developing countries and small island developing States, as well as the specific challenges facing the middle-income countries. Countries in situations of conflict also need special attention.

57. We recognize that baseline data for several of the targets remain unavailable, and we call for increased support for strengthening data collection and capacity building in Member States, to develop national and global baselines where they do not yet exist. We commit to addressing this gap in data collection so as to better inform the measurement of progress, in particular for those targets below which do not have clear numerical targets.

58. We encourage ongoing efforts by states in other fora to address key issues which pose potential challenges to the implementation of our Agenda; and we respect the independent mandates of those processes. We intend that the Agenda and its implementation would support, and be without prejudice to, those other processes and the decisions taken therein.

59. We recognise that there are different approaches, visions, models and tools available to each country, in accordance with its national circumstances and priorities, to achieve

---

① Contained in the report of the Open Working Group of the General Assembly on Sustainable Development Goals (A/68/970 and Corr.1; see also A/68/970/Add.1–3).

sustainable development; and we reaffirm that planet Earth and its ecosystems are our common home and that 'Mother Earth' is a common expression in a number of countries and regions.

Sustainable Development Goals

Goal 1.　End poverty in all its forms everywhere.

Goal 2.　End hunger, achieve food security and improved nutrition and promote sustainable agriculture.

Goal 3.　Ensure healthy lives and promote well-being for all at all ages.

Goal 4.　Ensure inclusive and equitable quality education and promote lifelong learning opportunities for all.

Goal 5.　Achieve gender equality and empower all women and girls.

Goal 6.　Ensure availability and sustainable management of water and sanitation for all.

Goal 7　Ensure access to affordable, reliable, sustainable and modern energy for all.

Goal 8.　Promote sustained, inclusive and sustainable economic growth, full and productive employment and decent work for all.

Goal 9.　Build resilient infrastructure, promote inclusive and sustainable industrialization and foster innovation.

Goal 10.　Reduce inequality within and among countries.

Goal 11.　Make cities and human settlements inclusive, safe, resilient and sustainable.

Goal 12.　Ensure sustainable consumption and production patterns.

Goal 13.　Take urgent action to combat climate change and its impacts[*].

Goal 14.　Conserve and sustainably use the oceans, seas and marine resources for sustainable development.

Goal 15.　Protect, restore and promote sustainable use of terrestrial ecosystems, sustainably manage forests, combat desertification, and halt and reverse land degradation and halt biodiversity loss.

Goal 16.　Promote peaceful and inclusive societies for sustainable development, provide access to justice for all and build effective, accountable and inclusive institutions at all levels.

Goal 17.　Strengthen the means of implementation and revitalize the Global Partnership for Sustainable Development.

* Acknowledging that the *United Nations Framework Convention on Climate Change* is the primary international, intergovernmental forum for negotiating the global response to climate change.

Notes

1. the *Universal Declaration of Human Rights*

《世界人权宣言》（简称《宣言》）是联合国的基本法之一，由 1948 年 12 月 10 日联合国大会通过第 217A（Ⅱ）号决议并颁布。这一具有历史意义的《宣言》颁布后，大会要求所有会员国广为宣传，并且"不分国家或领土的政治地位，主要在各级学校和其他教育机构加以传播、展示、阅读和阐述"。作为第一个人权问题的国际文件，《世界人权宣言》为国际人权领域的实践奠定了基础，对后来世界人民争取、维护、改善和发展自己的人权产生了深远影响。

随着世界非殖民化运动的蓬勃发展，在社会主义国家和许多取得独立的新兴国家的努力下，联合国于 1960 年通过了《给予殖民地国家和人民独立宣言》，1963 年和 1965 年先后通过了《消除一切形式种族歧视宣言》和《消除一切形式种族歧视国际公约》，并终于突破西方传统的人权概念，在 1966 年通过的《公民和政治权利国际公约》的第一条上明确规定："所有人民都有自决权。"之后，1968 年纪念《世界人权宣言》发表 20 周年的国际人权会议通过的《德黑兰宣言》，1977 年联合国通过的《关于人权新概念的决议》则以更加强烈的言辞谴责种族隔离、种族歧视和殖民主义，认为这些都是大规模侵犯人权的罪恶，消除它们是人类刻不容缓的最迫切的任务。应该说这些是对《宣言》的修正和发展。

2. the *Addis Ababa Action Agenda*

2015 年 7 月 15 日，在埃塞俄比亚首都亚的斯亚贝巴进行的联合国第三次发展筹资问题国际会议取得历史性突破。来自 193 个联合国会员国的与会代表经过谈判，正式就大会成果文件《亚的斯亚贝巴行动议程》达成一致，其中包括一系列旨在彻底改革全球金融实践并为解决经济、社会和环境挑战而创造投资的大胆措施。

经过联合国会员国长达数月的谈判，《亚的斯亚贝巴行动议程》正式通过，标志着在促进普遍和包容性的经济繁荣、提高人民福祉的同时保护环境方面加强全球合作伙伴关系的一个新的里程碑。出席大会的时任联合国秘书长潘基文对这一重要进展表示欢迎。他说，这项协议是为所有人建设一个可持续发展的未来进程中向前迈出的关键一步，它提供了资助可持续发展的全球框架。潘基文表示，此次亚的斯亚贝巴会议所取得的成果将为振兴可持续发展全球伙伴关系、确保实现"不让任何一个人被全球发展落下"的目标奠定重要的基础。

3. the *Beijing Declaration and Platform for Action*

1995 年，在第四次世界妇女大会中，来自 189 个国家的代表通过了《北京宣言和行动纲领》，获得了全世界的积极反响和拥护。《北京宣言和行动纲领》是指导全球妇女事业前进的重要指导性文件，标志着全球对妇女重要作用的觉醒，确定了各国和国际社会在提高妇女地位方面的共同目标。

世界妇女组织(World Women Organization)坚信妇女的全面发展是实现全人类和平、尊严和公平未来的基石。作为政府间组织，世界妇女组织与政府部门、非政府组织、民间团体、学术界和企业达成合作，推动世界各地的女性实现自身的全面发展。

目前，世界妇女组织总指挥部设于马来西亚吉隆坡，在纽约设有联络处，注册于联合国经济和社会事务部(UN DESA)。世界妇女组织认可并倡导联合国为实现一个和平、公正、可持续发展的全方位而努力。同时，联合国承认马来西亚是促进女童和女性参与科学领域的世界级领导者。

4. United Nations Conference on Housing and Sustainable Urban Development (Habitat Ⅲ)

第三届联合国住房和城市可持续发展大会（简称"人居三"）于 2016 年 10 月 17～

20 日在厄瓜多尔基多举行。联合国住房（人居）大会每 20 年召开一次（1976、1996 和 2016）。联合国大会在其第 66/207 号决议中决定召开人居三会议。人居三会议重振了全球对可持续城市化的承诺，并集中关注《新城市议程》的实施。

人居三会议是通过《2015 年后发展议程》后的首批联合国全球峰会之一。它为讨论重要的城市挑战和问题提供了机会，例如，如何规划和管理城市、镇和乡村以实现可持续发展。

Key Words and Phrases

1. aggregate	/'æɡrɪɡət/	adj.	an aggregate amount or score is made up of several smaller amounts or scores added together. 累计的
2. aquifer	/'ækwɪfə(r)/	n.	in geology, an aquifer is an area of rock underneath the surface of the earth which absorbs and holds water. 地下蓄水层；砂石含水层
3. auspices	/'ɔːspɪsɪz/	n.	under the auspices of sb./sth. 在某人/某物的帮助或支持下
4. biodiversity	/ˌbaɪəʊdaɪ'vɜːsəti/	n.	the diversity of plant and animal life in a particular habitat (or in the world as a whole) 生物多样性
5. commit	/kə'mɪt/	vt. & vi.	if you commit yourself to something, you say that you will definitely do it. 使（自己）承诺于
6. consultation	/ˌkɒns(ə)l'teɪʃ(ə)n/	n.	a consultation is a meeting to discuss something. Consultation is discussion about something. 磋商会议；磋商
7. desertification	/dɪˌzɜːtɪfɪ'keɪʃn/	n.	the gradual transformation of habitable land into desert; desertification is usually caused by climate change or by destructive use of the land. （土地）荒漠化
8. dimension	/daɪ'menʃ(ə)n/	n.	a particular dimension of something is a particular aspect of it. 方面
9. disparity	/dɪ'spærəti/	n.	if there is a disparity between two or more things, there is a noticeable difference between them. 明显差异

10. envisage	/ɪnˈvɪzɪdʒ/	vt.	if you envisage something, you imagine that it is true, real, or likely to happen. 设想；想象
11. endeavor	/ɪnˈdevə(r)/	vi.	attempt by employing effort 努力，尽力
12. eradicate	/ɪˈrædɪkeɪt/	vt.	to eradicate something means to get rid of it completely 根除，根绝；灭绝
13. hazardous	/ˈhæzədəs/	adj.	something that is hazardous is dangerous, especially to people's health or safety. 有危害的；冒险的
14. hygiene	/ˈhaɪdʒiːn/	n.	a condition promoting sanitary practices 卫生条件
15. indicator	/ˈɪndɪkeɪtə(r)/	n.	an indicator is a measurement or value that gives you an idea of what something is like. 指标；指示物
16. inequality	/ˌɪnɪˈkwɒləti/	n.	inequality is the difference in social status, wealth, or opportunity between people or groups. 不平等；不公平
17. low-lying	/ˌləʊ ˈlaɪɪŋ/	adj.	lying below the normal level 低于标准的
18. philanthropic	/ˌfɪlənˈθrɒpɪk/	adj.	of or relating to or characterized by philanthropy 慈善的
19. pledge	/pledʒ/	vt.	when someone pledges to do something, they promise in a serious way to do it. 保证（做某事）
20. premature	/ˈpremətʃə(r)/	adj.	born after a gestation period of less than the normal time 早产的
21. promulgate	/ˈprɒm(ə)lɡeɪt/	vt.	if people promulgate a new law or a new idea, they make it widely known. 散布；传播
22. reaffirm	/ˌriːəˈfɜːm/	vt.	if you reaffirm something, you state it again clearly and firmly. 重申
23. resilient	/rɪˈzɪliənt/	adj.	people and things that are resilient are able to recover easily and quickly from unpleasant or damaging events. 可迅速恢复的
24. revitalize	/ˌriːˈvaɪtəlaɪz/	vt.	to revitalize something that has lost its activity or its health means to make it

			active or healthy again. 使恢复元气；使复苏
25. segment	/'segmənt/	n.	a segment of something is one part of it, considered separately from the rest. 部分
26. spiraling	/'spaɪrəlɪŋ/	adj.	in the shape of a coil 盘旋的
27. stakeholder	/'steɪkhəʊldə(r)/	adj.	stakeholders are people who have an interest in a company's or organization's affairs. 股东；利益相关者
28. synthesis	/'sɪnθəsɪs/	n.	a synthesis of different ideas or styles is a mixture or combination of these ideas or styles. 结合体
29. tertiary	/'tɜːʃəri/	adj.	tertiary means third in order, third in importance, or at a third stage of development. 第三的
30. tyranny	/'tɪrəni/	n.	a tyranny is a cruel, harsh, and unfair government in which a person or small group of people have power over everyone else. 暴政；专横
31. anti-microbial resistance			抗生素耐药性
32. demographic dividend			人口红利
33. embark on sth.			开始着手做某事
34. give rise to			引起
35. gross domestic product			国内生产总值（简称 GDP）
36. ocean acidification			海洋酸化
37. spiraling conflict			不断升级的冲突

Exercises

Exercise 1　Reading Comprehension

Directions: *Read the Preamble of the text, and decide whether the following statements are true or false. Write T for true or F for false in the brackets in front of each statement.*

1. (　　) The biggest global challenge and an indispensable requirement for a more sustainable future for all people and the world is achieving gender equality and the empowerment of all women and girls.

2. (　　) A collection of 17 interlinked global goals and 169 targets was designed as a new agenda to complete the unfinished business of the Millennium Development Goals.

3. (　　) The Sustainable Development Goals are a bold commitment within all

countries to bring underprivileged people out of poverty and hunger, in all their forms and dimensions by 2030.

4. (   ) We can not hope for sustainable development without peace.

5. (   ) The participation of all developed countries, all stakeholders and enterprises has a more essential role in implementing the 2030 Agenda.

6. (   ) The integration of all the 17 Sustainable Development Goals contributes substantially to the attainment of the New Agenda.

Exercise 2   Skimming and Scanning

Directions: *Read the following passage excerpted from* ***Transforming Our World: the 2030 Agenda for Sustainable Development****. At the end of the passage, there are six statements. Each statement contains information given in one of the paragraphs of the passage. Identify the paragraph from which the information is derived. Each paragraph is marked with a letter. You may choose a paragraph more than once. Answer the questions by writing the corresponding letter in the brackets in front of each statement.*

A) This is an Agenda of unprecedented scope and significance. It is accepted by all countries and is applicable to all, taking into account different national realities, capacities and levels of development and respecting national policies and priorities.

B) We pledge to foster inter-cultural understanding, tolerance, mutual respect and an ethic of global citizenship and shared responsibility. We acknowledge the natural and cultural diversity of the world and recognize that all cultures and civilizations can contribute to, and are crucial enablers of, sustainable development.

C) All forms of discrimination and violence against women and girls will be eliminated, including through the engagement of men and boys.

D) We are committed to the prevention and treatment of non-communicable diseases, including behavioural, developmental and neurological disorders, which constitute a major challenge for sustainable development.

E) Never before have world leaders pledged common action and endeavour across such a broad and universal policy agenda. We are setting out together on the path towards sustainable development, devoting ourselves collectively to the pursuit of global development and of "win-win" cooperation which can bring huge gains to all countries and all parts of the world.

F) We will equally accelerate the pace of progress made in fighting malaria, HIV/AIDS, tuberculosis, hepatitis, Ebola and other communicable diseases and epidemics, including by addressing growing anti-microbial resistance and the problem of unattended diseases affecting developing countries.

G) We commit to making fundamental changes in the way that our societies produce and consume goods and services.

H) Women and girls must enjoy equal access to quality education, economic resources and political participation as well as equal opportunities with men and boys for employment, leadership and decision-making at all levels.

第 6 章　改变我们的世界：2030 年可持续发展议程

I) By 2030, reduce by one third premature mortality from non-communicable diseases through prevention and treatment and promote mental health and wellbeing.

1. (　　) The promotion of inter-cultural understanding and mutual respect among diverse cultures plays a vital role in the sustainable development.

2. (　　) The mission of the Agenda is to ensure, by 2030 and beyond, the full benefits to all people.

3. (　　) The *2030 Agenda* ensures all females' full and equal rights in education and job opportunity.

4. (　　) Certain health-related issues will be tackled in order to have sufficient acceleration of progress to combat all forms of transmissible diseases.

5. (　　) The *2030 Agenda* pledges to change unsustainable consumption and production patterns to more sustainable ones.

6. (　　) The *2030 Agenda* involves the entire world, developed and developing countries alike.

Exercise 3　Word Formation

Directions: *In this section, there are ten sentences from the text* **Transforming Our World: the 2030 Agenda for Sustainable Development**. *You are required to complete these sentences with the proper form of the words given in blanks.*

1. We are determined to protect the planet from degradation, including through sustainable ＿＿＿＿＿＿＿＿＿＿ and production, sustainably managing its natural resources and taking urgent action on climate change, so that it can support the needs of the present and future generations. (consume)

2. We are determined to ＿＿＿＿＿＿＿＿＿＿ the means required to implement this Agenda through a revitalized Global Partnership for Sustainable Development, based on a spirit of strengthened global solidarity, focused in particular on the needs of the poorest and most vulnerable and with the participation of all countries, all stakeholders and all people. (mobile)

3. The Goals and targets are the result of over two years of intensive public consultation and ＿＿＿＿＿＿＿＿＿＿ with civil society and other stakeholders around the world, which paid particular attention to the voices of the poorest and most vulnerable. (engage)

4. There are rising ＿＿＿＿＿＿＿＿＿＿ within and among countries. There are enormous disparities of opportunity, wealth and power. (equal)

5. Natural resource depletion and adverse impacts of environmental degradation, including desertification, drought, land degradation, freshwater ＿＿＿＿＿＿＿＿＿＿ and loss of biodiversity, add to and exacerbate the list of challenges which humanity faces. (scarce)

6. People who are vulnerable must be empowered. Those whose needs are reflected in the Agenda include all children, youth, persons with ＿＿＿＿＿＿＿＿＿＿ (of whom more than 80% live in poverty), people living with HIV/AIDS, older persons, indigenous peoples, refugees and internally displaced persons and migrants. (able)

7. We will devote resources to developing rural areas and sustainable agriculture and _____, supporting smallholder farmers, especially women farmers, herders and fishers in developing countries, particularly less developed countries. (fish)

8. We acknowledge the role of the diverse private sector, ranging from micro-enterprises to _____ to multinationals, and that of civil society organizations and philanthropic organizations in the implementation of the new Agenda. (cooperate)

9. Each government will also decide how these _____ and global targets should be incorporated in national planning processes, policies and strategies. (aspire)

10. We recognise that there are different approaches, visions, models and tools available to each country, in _____ with its national circumstances and priorities, to achieve sustainable development. (accord)

Exercise 4　Translation

Section A

Directions: *Read Transforming Our World: the 2030 Agenda for Sustainable Development, and complete the sentences by translating into English the Chinese given in blanks.*

1. Alongside continuing development priorities such as _____ _____ （消除贫困，健康，教育和食品安全与营养）, it sets out a wide range of economic, social and environmental objectives.

2. We resolve to take further effective measures and actions, in conformity with international law, to remove obstacles and constraints, strengthen support and meet the special needs of people living in areas affected by complex _____ _____ （人道主义紧急情况）) and in areas affected by terrorism.

3. We are also determined to promote sustainable tourism, to tackle water scarcity and water pollution, to strengthen cooperation on desertification, dust storms, land degradation and drought and to promote _____（灾后恢复能力和减少灾害风险）.

4. We will also take account of _____（人口趋势和人口预测） in our national rural and urban development strategies and policies.

5. It will facilitate an _____（全球高度参与）in support of implementation of all the Goals and targets, bringing together Governments, the private sector, civil society, the United Nations system and other actors and mobilizing all available resources.

6. We will _____（在国际间展开行动，确保安全、有序的定期移民） involving full respect for human rights and the humane treatment of migrants regardless of migration status, of refugees and of displaced persons.

7. We recognize the growing contribution of sport to the realization of development and peace in its promotion of tolerance and respect and the contributions it makes to_____ _____（增强妇女和青年、个人和社区的职能）as well as to health, education and social inclusion objectives.

8. Quality, accessible, timely and reliable disaggregated data will be needed to help with

the measurement of progress and to _____ （不让任何一个人掉队）.

9. We resolve to build a better future for all people, including the millions who have been denied the chance to lead decent, dignified and rewarding lives and to _____ _____（充分发挥潜力）.

10. We underline the important role and comparative advantage of _____ _____（一个资源充足、切合实际、协调一致、高效率和高成效的联合国系统）in supporting the achievement of the SDGs and sustainable development.

Section B

Directions: *Translate the following sentences from English into Chinese.*

1. The 17 Sustainable Development Goals and 169 targets which we are announcing today demonstrate the scale and ambition of this new universal Agenda. They seek to build on the Millennium Development Goals and complete what these did not achieve. They seek to realize the human rights of all and to achieve gender equality and the empowerment of all women and girls. They are integrated and indivisible and balance the three dimensions of sustainable development: the economic, social and environmental. (*Preamble*)

2. The interlinkages and integrated nature of the Sustainable Development Goals are of crucial importance in ensuring that the purpose of the new Agenda is realised. If we realize our ambitions across the full extent of the Agenda, the lives of all will be profoundly improved and our world will be transformed for the better. (*Preamble*)

3. We resolve, between now and 2030, to end poverty and hunger everywhere; to combat inequalities within and among countries; to build peaceful, just and inclusive societies; to protect human rights and promote gender equality and the empowerment of women and girls; and to ensure the lasting protection of the planet and its natural resources. We resolve also to create conditions for sustainable, inclusive and sustained economic growth, shared prosperity and decent work for all, taking into account different levels of national development and capacities. (*Declaration*)

4. The challenges and commitments contained in these major conferences and summits are interrelated and call for integrated solutions. To address them effectively, a new approach is needed. Sustainable development recognizes that eradicating poverty in all its forms and dimensions, combatting inequality within and among countries, preserving the planet, creating sustained, inclusive and sustainable economic growth and fostering social inclusion are linked to each other and are interdependent. (*Declaration*)

5. We recognize that each country has primary responsibility for its own economic and social development. The new Agenda deals with the means required for implementation of the Goals and targets. We recognize that these will include the mobilization of financial resources as well as capacity-building and the transfer of environmentally sound technologies to developing countries on favourable terms, including on concessional and preferential terms, as mutually agreed. Public finance, both domestic and international, will play a vital role in providing essential services and public goods and in catalyzing other sources of finance.

(*Means of Implementation*)

6. The SDGs and targets are integrated and indivisible, global in nature and universally applicable, taking into account different national realities, capacities and levels of development and respecting national policies and priorities. Targets are defined as aspirational and global, with each government setting its own national targets guided by the global level of ambition but taking into account national circumstances. Each government will also decide how these aspirational and global targets should be incorporated in national planning processes, policies and strategies. (*Sustainable Development Goals and targets*)

Extensive Readings

Passage 1

Directions: *Read the following passage and choose the best answer for each of the following questions according to the information given in the passage.*

Over the past five years, under the strong leadership of President Xi Jinping, China has put people's well-being front and center, followed a new vision of innovative, coordinated, green, open and shared development, and taken comprehensive measures to implement the *2030 Agenda* according to its national implementation plan that was among the first of the kind in the world, and through an inter-ministerial coordination mechanism made up of 45 ministries and agencies, integrating its implementation efforts with the country's 13th Five-Year Plan and other medium-to-long-term development strategies. China has also actively participated in international development cooperation to promote global implementation endeavors. All these efforts have yielded remarkable results.

China has eradicated extreme poverty, and ensured food security. At the end of 2020, China won its fight against poverty as scheduled, ending poverty for the 98.99 million rural residents living below the current poverty line, and meeting the targets of SDG 1 ten years ahead of schedule. Building on these achievements, China is prioritizing the development of agriculture and rural areas, and implementing the rural revitalization strategy in a comprehensive manner. In 2020, China reaped bumper harvest in grain production for the 17th year in a row. The "rice bowl" of the Chinese people is tightly held in the hands of the Chinese themselves.

China has actively taken climate actions, and contributed to global green development. Following the philosophy that lucid water and lush mountain are <u>invaluable</u> assets, China has accelerated the transition towards green development. China has won the three critical battles to keep its sky blue, water clear, and soil pollution-free and made remarkable headway in holistic protection and treatment of its mountains, rivers, forests, farmland, lakes, grassland and deserts. China has firmly implemented the *Paris Agreement* and actively participated in global climate governance. Carbon intensity in China has accumulatively dropped by 18.8%. Clean energy now accounts for 23.4% of China's energy mix. China leads the world in installed capacity and output of photovoltaic and wind power generation. China has scaled up its nationally determined contributions and aims to peak its carbon dioxide emissions before 2030 and achieve carbon neutrality before 2060, injecting strong impetus into global climate actions as well as green and low-carbon development worldwide.

China has met the challenge of the pandemic with success and improved public health governance. Putting people and their lives above everything else, China has scored major strategic achievements in fighting COVID-19 and protected people's rights to life and health. China has increased investment in public health infrastructure and woven the world's largest social safety net with basic medical insurance covering more than 1.3 billion people. With upgraded public health services, Chinese people's sense of fulfillment, happiness and security is steadily increased.

China has shouldered its responsibilities as a major country and advanced international development cooperation. China champions mankind's common values of peace, development, fairness, justice, democracy and freedom, and strives to build a community of shared future for mankind. It has stepped up efforts to promote synergy between the Belt and Road Initiative and the *2030 Agenda*, promoted the green BRI, deepened South-South cooperation, and helped other developing countries implement the *2030 Agenda* to the best of its capability. China has provided assistance to more than 160 countries and international organizations in combating COVID-19, and donated and exported pandemic response supplies to over 200 countries and regions. China has joined the COVAX, provided or is providing vaccine assistance to more than 80 developing countries most in need, and exported vaccines to more than 40 countries, making its due contribution to the global fight against the pandemic.

Likewise, the global implementation of the *2030 Agenda* calls for strong political leadership, effective institutional guarantees, scientific and technological innovation, extensive social mobilization and pragmatic global partnership. The COVID-19 pandemic has dealt a severe blow to the global implementation of the *2030 Agenda*. China stands ready to work with other countries to consolidate political will, put development first, strengthen the means of implementation, take joint actions and address special difficulties of developing countries to leave no one and no country behind.

1. Which of the following is **NOT** one of the efforts made by China to fully implement the *2030 Agenda*?_____

  A. China has been actively engaging with international development cooperation.

  B. China has integrated efforts from the 13th Five-Year Plan to serve better for global goals.

  C. China has launched effective development strategies to provide significant support.

  D. China has stressed its leadership and responsibilities to promote a new vision.

2. The author mentioned "the 'rice bowl' of the Chinese people" in Paragraph 2 to _____.

  A. present that China has enhanced its agricultural productive capacity

B. prove that China has made great progress in ending hunger

C. prove that every Chinese has his own "rice bowl"

D. show that there is no more hunger in every corner of China

3. What does the underlined word "invaluable" probably mean in Paragraph 3?_____

   A. valueless   B. very valuable   C. worthless   D. too expensive

4. Which of the following is **NOT** one of the battles that China has won to contribute to global green development?_____

   A. lowering the carbon intensity

   B. expanding the use of clean energy

   C. withdrawing from the *Paris Agreement*

   D. offering treatment for polluted mountains, rivers, forests, etc.

5. According to Paragraphs 5 and 6, which of the following is **NOT** true?_____

   A. China has made vigorous efforts to align the Belt and Road Initiative with the *2030 Agenda*.

   B. China has done everything in its power to keep the virus away from its people and people outside China.

   C. Over 200 countries and regions have purchased pandemic response supplies exported from China.

   D. The COVID-19 pandemic presents an enormous challenge for reaching the *2030 Agenda*.

Passage 2

Directions: *In this section, there is a passage with twelve blanks. You are required to select one word for each blank from a list of choices given in a word bank following the passage. Read the passage through carefully before making your choices. Each choice in the bank is identified by a letter. You may not use any of the words in the bank more than once.*

A. undermine	B. framework	C. combat	D. temperature
E. anticipated	F. survival	G. commitment	H. promotion
I. concern	J. necessities	K. associated	L. cooperation

For many, a warming climatic system is expected to impact the availability of basic __1__ like freshwater, food security, and energy, while efforts to redress climate change, both through adaptation and mitigation, will similarly inform and shape the global development agenda. The links between climate change and sustainable development are strong. Poor and developing countries, particularly least developed countries, will be among those most adversely affected and least able to cope with the __2__ shocks to their social, economic and natural systems.

The international political response to climate change began at the Rio Earth Summit in 1992, where the '*Rio Convention*' included the adoption of the *UN Framework Convention on Climate Change* (UNFCCC). This convention set out a __3__ for action aimed at stabilizing atmospheric concentrations of greenhouse gases (GHGs) to avoid "dangerous anthropogenic

Chapter 6 Transforming Our World: the 2030 Agenda for Sustainable Development

interference with the climate system." The UNFCCC which entered into force on 21 March 1994, now has a near-universal membership of 197 parties. In December 2015, the 21st Session of the Conference of the Parties (COP21/CMP1) convened in Paris, France, and adopted the *Paris Agreement*, a universal agreement which aims to keep a global __4__ rise for this century well below 2 degrees Celsius, with the goal of driving efforts to limit the temperature rise to 1.5 degrees Celsius above pre-industrial levels.

In the *2030 Agenda* for Sustainable Development, Member States express their __5__ to protect the planet from degradation and take urgent action on climate change. The Agenda also identifies, in its paragraph 14, climate change as "one of the greatest challenges of our time" and worries about "its adverse impacts __6__ the ability of all countries to achieve sustainable development. Increases in global temperature, sea level rise, ocean acidification and other climate change impacts are seriously affecting coastal areas and low-lying coastal countries, including many least developed countries and Small Island Developing States. The __7__ of many societies, and of the biological support systems of the planet, is at risk".

Sustainable Development Goal 13 aims to "take urgent action to __8__ climate change and its impact", while acknowledging that the *United Nations Framework Convention on Climate Change* is the primary international, intergovernmental forum for negotiating the global response to climate change.

More specifically, the __9__ targets of SDG 13 focus on the integration of climate change measures into national policies, the improvement of education, awareness-raising and institutional capacity on climate change mitigation, adaptation, impact reduction and early warnings. SDG 13's alphabetical targets also call for the implementation of the commitment undertaken at the UNFCCC and for the __10__ of mechanisms able to increase capacity for effective climate change related planning and management in least developed countries and Small Island Developing States.

The outcome document of the Rio+20 Conference, *The Future We Want*, underscores climate change as "an inevitable and urgent global challenge with long-term implications for the sustainable development of all countries". Through the document, Member States express their __11__ about the continuous rising of emissions of greenhouse gases and the vulnerability of all countries, particularly developing countries, to the adverse impacts of climate change. Given these concerns, Member States have called for the widest __12__ and participation of all countries in an effective and appropriate international response to climate change.

Further Studies and Post-Reading Discussion

Task 1
Directions: *Surf the Internet and find more information about* **Transforming Our World: the 2030 Agenda for Sustainable Development**. *Work in groups and work out a report on one of the following topics.*

1. Purposes of *Transforming Our World: the 2030 Agenda for Sustainable Development*

(the *2030 Agenda*).
2. China's national plan in implementing the *2030 Agenda*.
3. China's contributions to the global sustainable development.

Task 2
Directions: *Read the following sentences on Eco-Civilization and make a speech on your understanding of the eco-environmental conservation.*

碳达峰碳中和

积极稳妥推进碳达峰碳中和。实现碳达峰碳中和是一场广泛而深刻的经济社会系统性变革。立足我国能源资源禀赋，坚持先立后破，有计划分步骤实施碳达峰行动。完善能源消耗总量和强度调控，重点控制化石能源消费，逐步转向碳排放总量和强度"双控"制度。 推动能源清洁低碳高效利用，推进工业、建筑、交通等领域清洁低碳转型。深入推进能源革命，加强煤炭清洁高效利用，加大油气资源勘探开发和增储上产力度，加快规划建设新型能源体系，统筹水电开发和生态保护，积极安全有序发展核电，加强能源产供储销体系建设，确保能源安全。 完善碳排放统计核算制度，健全碳排放权市场交易制度。提升生态系统碳汇能力。积极参与应对气候变化全球治理。（摘自习近平《在中国共产党第二十次全国代表大会上的报告》）	Working actively and prudently toward the goals of reaching peak carbon emissions and carbon neutrality. Reaching peak carbon emissions and achieving carbon neutrality will mean a broad and profound systemic socio-economic transformation. Based on China's energy and resource endowment, we will advance initiatives to reach peak carbon emissions in a well-planned and phased way in line with the principle of building the new before discarding the old. We will exercise better control over the amount and intensity of energy consumption, particularly of fossil fuels, and transition gradually toward controlling both the amount and intensity of carbon emissions. We will promote clean, low-carbon, and high-efficiency energy use and push forward the clean and low-carbon transition in industry, construction, transportation, and other sectors. We will thoroughly advance the energy revolution. Coal will be used in a cleaner and more efficient way, and greater efforts will be made to explore and develop petroleum and natural gas, discover more untapped reserves, and increase production. We will speed up the planning and development of a system for new energy sources, properly balance hydropower development and ecological conservation, and develop nuclear power in an active, safe, and orderly manner. We will strengthen our systems for energy production, supply, storage, and marketing to ensure energy security. We will improve the statistics and accounting system and the cap-and-trade system for carbon emissions. The carbon absorption capacity of ecosystems will be boosted. We will get actively involved in global governance in response to climate change. (Excerpt from *Report to the 20th National Congress of the Communist Party of China* by Xi Jinping)

第 7 章 2015 年后国际森林安排决议

Chapter 7　International Arrangement on Forests beyond 2015

Background and Significance

2000 年 10 月，联合国经济及社会理事会成立了联合国森林论坛（United Nations Forest Forum, UNFF）。2015 年 5 月，在 UNFF 第 11 届会议上，各成员国通过了具有里程碑意义的《2015 年后国际森林安排决议》（*International Arrangement on Forests beyond 2015*，以下简称《决议》）。《决议》全面细致地对 2015 年至 2030 年的国际森林安排做出了明确阐述，决定了未来 15 年全球森林政策走向和全球林业可持续发展战略，对提高森林在全球可持续发展中的战略地位有着重要意义。根据《决议》，新国际森林安排包括 UNFF 及其成员国、UNFF 秘书处、森林合作伙伴关系、全球森林资金网络和 UNFF 信托基金，并欢迎国际、区域和次区域组织和进程、主要群体和其他利益攸关方参与，有效期至 2030 年。国际森林安排的主要目标是：履行《联合国森林文书》，推动全球森林可持续经营；加强森林对 2015 年后发展议程做出的贡献；促进国际合作，包括南北合作、南南合作、三边合作，以及各级公私伙伴关系和跨部门合作。《决议》强调，2015 年后国际森林安排应以透明、有效、高效和负责任的方式运行，且应为其他涉林协议、进程和倡议提供附加值，并促进国际合作与协同增效。

Text Study

International Arrangement on Forests beyond 2015

The Economic and Social Council,

Recalling its resolution 2000/35 of 18 October 2000, by which it established the international arrangement on forests,

Recalling also the principles set out in the *Rio Declaration on Environment and Development*,[①] and recalling the outcome document of the United Nations Conference on

---

① Report of the United Nations Conference on Environment and Development, Rio de Janeiro, 3–14 June 1992, vol. I, Resolutions Adopted by the Conference (United Nations publication, Sales No. E.93.I.8 and corrigendum), resolution 1, annex I.

Sustainable Development, entitled "*The Future We Want*",①

Recalling further its resolution 2006/49 of 28 July 2006 and United Nations Forum on Forests resolution 10/2 of 19 April 2013,② providing for the review in 2015 of the effectiveness of the international arrangement on forests, including its scope and its preparatory process,

Recognizing the achievements of the international arrangement on forests since its inception, in particular the adoption by the General Assembly of the non-legally binding instrument on all types of forests③ adopted by the Forum, as well as the contributions of the Collaborative Partnership on Forests, while stressing the continued challenges of and the need to strengthen the international arrangement on forests and to continue efforts to contribute to the promotion and implementation of sustainable forest management,

Acknowledging the progress made by countries and stakeholders towards sustainable forest management, including the implementation of the non-legally binding instrument on all types of forests and the achievement of its global objectives on forests at the local, national, regional and international levels, taking into account different visions, approaches, models and tools to achieve sustainable development,

Welcoming the significant forest-related developments in other forums, in particular in the context of the *Rio Conventions*④, their continued contribution to sustainable forest management and the importance of cooperation and synergies between these forums and the international arrangement on forests,

Welcoming also the recognition given to forests and sustainable forest management by the Open Working Group of the General Assembly on Sustainable Development Goals in its proposed sustainable development goals and associated targets, and emphasizing the economic, social and environmental contributions of all types of forests to the achievement of the post-2015 development agenda,

Noting the contributions made by countries, organizations and other stakeholders to the review of the international arrangement on forests, including the views submitted by Member States of the Forum and major groups and the reports of the independent assessment of the international arrangement on forests, the Open-ended Intergovernmental Ad Hoc Expert Group on the International Arrangement on Forests and the initiatives

---

① General Assembly resolution 66/288, annex.

② See Official Records of the Economic and Social Council, 2013, Supplement No. 22 (E/2013/42), chap. I, sect. B.

③ General Assembly resolution 62/98, annex.

④ *Convention on Biological Diversity* (United Nations, Treaty Series, vol. 1760, No. 30619), United Nations Convention to Combat Desertification in *Those Countries Experiencing Serious Drought and/or Desertification, Particularly in Africa* (United Nations, Treaty Series, vol. 1954, No. 33480) and United Nations Framework Convention on Climate Change (United Nations, Treaty Series, vol. 1771, No. 30822).

hosted by China, Nepal and Switzerland,

Stressing the need to strengthen the capacity of the international arrangement on forests to foster coherence on forest-related policies, catalyse the implementation of and financing for sustainable forest management, and promote coordination and collaboration on forest issues at all levels, as well as coherence between the international arrangement on forests and the post-2015 development agenda.

I
International Arrangement on Forests beyond 2015

1. *Decides:*

(a) To strengthen the international arrangement on forests and extend it to 2030;

(b) That the international arrangement on forests is composed of the United Nations Forum on Forests and its Member States, the secretariat of the Forum, the Collaborative Partnership on Forests, the Global Forest Financing Facilitation Network[①] and the Trust Fund for the United Nations Forum on Forests;

(c) That the international arrangement on forests involves as partners interested international, regional and subregional organizations and processes, major groups and other stakeholders;

(d) That the objectives of the international arrangement on forests are:

• To promote the implementation of sustainable management of all types of forests, in particular the implementation of the non-legally binding instrument on all types of forests;

• To enhance the contribution of all types of forests and trees outside forests to the post-2015 development agenda;

• To enhance cooperation, coordination, coherence and synergies on forest-related issues at all levels;

• To foster international cooperation, including North-South, South-South and triangular cooperation, as well as public-private partnerships and cross-sectoral cooperation at all levels;

• To support efforts to strengthen forest governance frameworks and means of implementation, in accordance with the non-legally binding instrument on all types of forests, in order to achieve sustainable forest management;

• To strengthen long-term political commitment to the achievement of the objectives listed in paragraph 1 (d) of the present resolution;

• That the international arrangement on forests beyond 2015 should operate in a transparent, effective, efficient and accountable manner and should provide added value and contribute to enhanced coherence, cooperation and synergies with respect to other

---

① See paragraph 13 (a) of the present resolution.

forest-related agreements, processes and initiatives;

2. *Emphasizes* that the objectives of the international arrangement on forests beyond 2015 should be achieved through the actions, individually and collectively, of Member States, international, regional and subregional organizations and processes, major groups and other stakeholders;

II
United Nations Forum on Forests beyond 2015

3. *Decides* that the core functions of the Forum are:

(a) To provide a coherent, open, transparent and participatory global platform for policy development, dialogue, cooperation and coordination on issues related to all types of forests, including emerging issues, in an integrated and holistic manner, including through cross- sectoral approaches;

(b) To promote, monitor and assess the implementation of sustainable forest management, in particular the non-legally binding instrument on all types of forests and the achievement of its global objectives on forests, and mobilize, catalyse and facilitate access to financial, technical and scientific resources to this end;

(c) To promote governance frameworks and enabling conditions at all levels to achieve sustainable forest management;

(d) To promote coherent and collaborative international policy development on issues related to all types of forests;

(e) To strengthen high-level political engagement, with the participation of major groups and other stakeholders, in support of sustainable forest management;

4. *Reaffirms* that, as set out in paragraph 4 of Economic and Social Council resolution 2000/35, the Forum is a subsidiary body of the Council with universal membership, which operates under the rules of procedure of the functional commissions and reports to the Council and, through the Council, to the General Assembly;

5. *Decides* that the Forum shall continue to operate according to the provisions specified in paragraphs 4 (a) to (e) of resolution 2000/35 unless otherwise provided in the present resolution;

6. *Also decides* to improve and strengthen the functioning of the Forum beyond 2015 by requesting the Forum:

(a) To carry out its core functions on the basis of a strategic plan for the period 2017—2030, as defined in "Strategic Plan" of the present resolution;

(b) To restructure its sessions and enhance its intersessional work to maximize the impact and relevance of its work, including by fostering an exchange of experiences and lessons learned among countries, regional, subregional and non-governmental partners and the Collaborative Partnership on Forests;

(c) To hold annual sessions of the Forum for a period of five days;

(d) To convene high-level segments not to exceed two days during sessions of the Forum, as required, to accelerate action towards sustainable forest management and address forest-related global challenges and emerging issues; such segments may include a partnership forum involving the heads of member organizations of the Collaborative Partnership on Forests and leaders from the private sector, philanthropic and civil society organizations and other major groups;

(e) To enhance the contributions to the work of the Forum by country-led and similar initiatives by ensuring that they directly support the priorities of the Forum as defined in its four-year work programmes and that their outcomes are considered by the Forum, and update the Forum guidelines in this regard;

(f) In line with paragraph 6 (b) of the present resolution, to dedicate the odd-year session of the Forum to discussions on implementation and technical advice for the purpose of focusing the attention of Member States on the specific tasks listed below; the summaries of these discussions, including possible proposals, will be reported to the Forum at its subsequent sessions in the even years for further consideration and recommendations. The specific tasks are:

• To assess the progress in, and make possible proposals on, the implementation of the resolutions and decisions of the Forum, the non-legally binding instrument on all types of forests and the strategic plan;

• To assess the progress in, and make possible proposals on, enhancing policy coherence, dialogue and cooperation on forests, fostering synergies in global forest-related processes and strengthening the common international understanding of the concept of sustainable forest management as set out in the non-legally binding instrument on all types of forests;

• To monitor and assess the work and the performance of the strengthened facilitative process;

• To review and advise on the availability of resources for sustainable forest management funding, including the strengthened facilitative process, and ensure that its operation is consistent with guidelines to be approved by the Forum;

• To review and make possible proposals on the operation of the Forum Trust Fund;

(g) In line with paragraph 6 (b) of the present resolution, to dedicate the odd-year sessions of the Forum:

• To serving as an opportunity for the Collaborative Partnership on Forests and its member organizations, regional and subregional organizations and processes, major groups and other relevant stakeholders to provide technical advice and input to the Forum;

• To facilitating the sharing of knowledge and best practices, including the science-policy interface;

III
Non-legally Binding Instrument on All Types of Forests beyond 2015

7. *Reaffirms* the continued validity and value of the non-legally binding instrument on all types of forests, including its global objectives on forests, and emphasizes the need to strengthen and catalyse its implementation at all levels, taking into account forest-related developments since 2007, including developments in the context of the *Rio Conventions*;[①]

8. *Decides* to extend the timeline of the global objectives on forests to 2030, in line with the post-2015 development agenda, and to rename the non-legally binding instrument on all types of forests the *"United Nations Forest Instrument"*, recognizing that the voluntary, non-binding character of the forest instrument, as set out in principle 2 (a) of the instrument, remains unchanged;

9. *Recommends* to the General Assembly that it adopt the modifications referred to in paragraph 8 of the present resolution during its seventieth session and not later than December 2015;

10. *Urges* Member States to utilize the non-legally binding instrument on all types of forests as an integrated framework for national action and international cooperation for implementing sustainable forest management and forest-related aspects of the post-2015 development agenda;

IV
Catalysing Financing for Implementation

11. *Reiterates* that there is no single solution to address all of the needs in terms of forest financing and that a combination of actions is required at all levels, by all stakeholders and from all sources, public and private, domestic and international, bilateral and multilateral;

12. *Welcomes* the positive work carried out by the facilitative process to date, and recognizes that it has yet to fulfil its potential as set out in the resolutions contained in the reports of the Forum on the special session of its ninth session[②] and on its ninth session;[③]

13. *Decides*, in order to strengthen and make the facilitative process more effective:

(a) That the name of the facilitative process shall be changed to the "Global Forest Financing Facilitation Network";

(b) To set clear priorities for the strengthened facilitative process in the strategic plan,

---

① See P. 152④.

② E/2009/118, sect. I.B.

③ Official Records of the Economic and Social Council, 2011, Supplement No. 22 (E/2011/42), chap. I, sect. B.

as described in "Strategic Plan" of the present resolution;

(c) That it should promote the design of national forest financing strategies to mobilize resources for sustainable forest management, including existing national initiatives, within the framework of national forest programmes or their equivalent, to facilitate access to existing and emerging financing mechanisms, including the Global Environment Facility and the Green Climate Fund, consistent with their mandates, in order to implement sustainable forest management;

(d) That it should serve as a clearing house on existing, new and emerging financing opportunities and as a tool for sharing lessons learned from successful projects, building on the Collaborative Partnership on Forests online sourcebook for forest financing;

(e) That it should ensure that special consideration is given to the special needs and circumstances of Africa, the least developed countries, low-forest-cover countries, high-forest-cover countries, medium-forest-cover low-deforestation countries and small island developing States, as well as countries with economies in transition, in gaining access to funds;

(f) To enhance the capacity of the secretariat to effectively and efficiently manage the strengthened facilitative process;

(g) To strengthen collaboration with the Collaborative Partnership on Forests in implementting the activities of the strengthened facilitative process;

14. *Also decides*, with the aim of strengthening the facilitative process:

(a) To request the secretariat, in consultation with the members of the Forum and the Collaborative Partnership on Forests, to make recommendations on ways to further increase the effectiveness and efficiency of the operation of the strengthened facilitative process and submit them for consideration by the Forum at its session in 2018;

(b) To welcome the report of the secretariat of the Global Environment Facility to the Forum on the mobilization of financial resources through the sustainable forest management/ REDD-plus incentive programme under the fifth replenishment of the Facility, and invite the secretariat of the Facility to periodically provide information on the mobilization of financial resources and funds that are dedicated to sustainable forest management;

(c) To also welcome the decision taken by the Assembly of the Global Environment Facility at its session in May 2014 to include a sustainable forest management strategy in the sixth replenishment period of the Facility (2014—2018) to support the sustainable management of all types of forests;

(d) To encourage eligible Member States, taking into account the cross-sectoral nature of sustainable forest management, to make full use of the existing potential of the sustainable forest management strategy under the sixth replenishment of the Global Environment Facility to harness synergies across the focal areas of the Facility in order to further reinforce the importance of sustainable forest management for integrating environmental and development aspirations;

(e) To invite the Council of the Global Environment Facility to request the secretariat of the Facility to discuss with the secretariat of the Forum arrangements to facilitate collaboration between the Facility and the Forum to support eligible countries in gaining access to funding for sustainable forest management;

(f) To request the secretariat of the Forum to engage in discussions with the secretariat of the Global Environment Facility in order to facilitate collaboration between the Facility and the Forum to support eligible countries in gaining access to funding for sustainable forest management, and to report to the Forum on this issue;

15. *Invites* the Global Environment Facility to consider:

(a) Options for establishing a new focal area on forests during the next replenishment of the Facility and continuing to seek to improve existing forest finance modalities;

(b) Designating among its staff a liaison to serve as a link between the Forum and the Facility, in order to facilitate access to funding for sustainable forest management;

V

Monitoring, Assessment and Reporting

16. *Decides:*

(a) To invite Member States to continue to monitor and assess progress towards implementing sustainable forest management, including the non-legally binding instrument on all types of forests and the global objectives on forests, and to submit on a voluntary basis national progress reports to the Forum;

(b) To note the ongoing efforts of the Collaborative Partnership on Forests and its members and other relevant entities and processes to work jointly to further streamline and harmonize reporting, reduce reporting burdens and synchronize data collection, taking into account the collaborative forest resources questionnaire developed as part of the Global Forest Resources Assessment 2015, in order to foster synergy and coherence;

(c) To request the secretariat of the Forum, in consultation with Member States, the Collaborative Partnership on Forests and its members and other relevant entities and processes, as well as criteria and indicators processes, to propose for consideration by the Forum at its next session a cycle and a format for national reporting and the enhancement of voluntary monitoring, assessment and reporting under the international arrangement on forests as part of the strategic plan referred to in "Strategic Plan" of the present resolution, taking into account and utilizing existing data collection mechanisms;

(d) To request the secretariat of the Forum to continue to make the reports on its sessions, as well as other relevant inputs, available to relevant United Nations bodies and other international forest-related organizations, instruments and intergovernmental processes;

VI

Secretariat of the Forum

17. *Decides* that the secretariat of the Forum:

(a) Should continue:
· To service and support the Forum, its Bureau and related intersessional activities, including by organizing and supporting meetings, providing operational and logistical support and preparing documentation;
· To administer the Forum Trust Fund consistent with guidance provided by the Forum, including regular and transparent reporting;
· To manage the strengthened facilitative process;
· To promote inter-agency collaboration, including by serving as a member of and providing secretariat services to the Collaborative Partnership on Forests;
· To provide, upon request, technical support to country-led initiatives and similar initiatives led by international, regional and subregional organizations and processes, and major groups in support of the priorities of the Forum;
· To liaise with and facilitate the participation and involvement of countries, organizations, major groups and other stakeholders in activities of the Forum, including intersessional activities;

(b) Should perform the following additional functions:
· Service and support the working group of the Forum, including by organizing and supporting meetings, providing operational and logistical support and preparing documentation;
· Manage the Global Forest Financing Facilitation Network and implement its activities in collaboration with relevant members of the Collaborative Partnership on Forests;
· Promote coherence, coordination and cooperation on forest-related issues, including by liaising with the secretariats of the *Rio Conventions*;
· Work within the United Nations system to support countries in aligning forests and the international arrangement on forests with their considerations on the post-2015 development agenda;

18. *Reaffirms* that the secretariat of the Forum continues to be located at United Nations Headquarters in New York;

19. *Recommends* to the General Assembly that it consider strengthening the secretariat of the Forum, taking into account the provisions of the present resolution;

VII

Collaborative Partnership on Forests

20. *Decides* that the core functions of the Collaborative Partnership on Forests are:
(a) To support the work of the Forum and its member countries;
(b) To provide scientific and technical advice to the Forum, including on emerging issues;
(c) To enhance coherence as well as policy and programme cooperation and coordination at all levels among its member organizations, including through joint programming and the submission of coordinated proposals to their respective governing bodies, consistent with

their mandates;

(d) To promote the implementation of the non-legally binding instrument on all types of forests, including the achievement of its global objectives on forests, and the contribution of forests to the post-2015 development agenda;

21. *Reaffirms* that the Collaborative Partnership on Forests should continue:

(a) To receive guidance from the Forum and submit coordinated inputs and progress reports to sessions of the Forum;

(b) To operate in an open, transparent and flexible manner;

(c) To undertake periodic reviews of its effectiveness;

22. *Encourages* the Collaborative Partnership on Forests and its member organizations:

(a) To strengthen the Partnership by formalizing its working modalities, including through consideration of a multilateral memorandum of understanding, and by developing procedures for its effective functioning and operation;

(b) To identify ways to stimulate broader participation by existing member organizations in its various activities;

(c) To assess its membership and the potential added value of additional members with significant forest-related expertise;

(d) To identify ways to actively involve major groups and other stakeholders in activities of the Partnership;

(e) To develop a workplan, aligned with the strategic plan referred to in "Strategic Plan" of the present resolution, to identify priorities for collective actions by all of the members of the Partnership or subsets of members and the resource implications of such actions;

(f) To prepare periodic reports on the Partnership activities, achievements and resource allocations suitable for a wide range of audiences, including potential donors;

(g) To further develop and expand its thematic joint initiatives, taking into account the strengths and focuses of the members of the Partnership;

23. *Invites* the governing bodies of member organizations of the Collaborative Partnership on Forests to include in their work programmes dedicated funding to support Partnership activities, as well as budgeted activities supporting the priorities of the Forum as outlined in the strategic plan referred to in "Strategic Plan" of the present resolution, consistent with their mandates;

24. *Calls* upon Member States, as well as other members of the governing bodies of member organizations of the Collaborative Partnership on Forests, to support the work of the Partnership, including by considering dedicated funding for Partnership activities consistent with the respective mandates of Partnership members as an essential strategy for improving cooperation, synergies and coherence on forest issues at all levels;

VIII
Regional and Subregional Involvement

25. *Requests* the Forum to strengthen its collaboration with relevant regional and subregional forest-related mechanisms, institutions and instruments, organizations and processes in order to facilitate the implementation of the non-legally binding instrument on all types of forests, including the achievement of its global objectives on forests, as well as to facilitate their inputs to sessions of the Forum;

26. *Requests* the secretariat of the Forum to consult with relevant regional and subregional forest-related mechanisms, institutions and instruments, organizations and processes on means to enhance collaboration between them and the Forum, including regarding the implementation of the strategic plan and the quadrennial programmes of work referred to in "Strategic Plan" of the present resolution;

27. *Invites* relevant regional and subregional mechanisms, institutions and instruments, organizations and processes in a position to do so to consider, consistent with their mandates, developing or strengthening programmes on sustainable forest management, including facilitating the implementation of the non-legally binding instrument on all types of forests and relevant aspects of the post-2015 development agenda, as well as to provide coordinated inputs and recommendations to sessions of the Forum;

28. *Invites* Member States to consider, on a voluntary basis and as appropriate, establishing or strengthening regional and subregional processes or platforms for forest policy development, dialogue and coordination to promote sustainable forest management while seeking to avoid fragmentation;

IX
Involvement of Major Groups and Other Stakeholders

29. *Recognizes* the importance of the continued and enhanced participation of major groups and other stakeholders in the sessions of the Forum and its intersessional activities;

30. *Decides*, in this regard, that the provisions of paragraphs 14 to 16 of General Assembly resolution 67/290 of 9 July 2013 apply mutatis mutandis to the Forum in view of the existing modalities and practices of the Forum;

31. *Invites* major groups and other stakeholders to enhance their contributions to the work of the international arrangement on forests beyond 2015;

32. *Invites* Member States to consider enhancing the participation and contributions of representatives of major groups and other stakeholders in country-led initiatives;

33. *Requests* the secretariat of the Forum to promote the involvement of major groups

and other stakeholders in the work of the Forum, in particular leaders from the private and non-governmental sectors, including forest industries, local communities and philanthropic organizations, and to enhance the interaction of the Forum with such stakeholders;

X
International Arrangement on Forests and the Post-2015 Development Agenda

34. *Stresses* the need to ensure coherence and consistency between the international arrangement on forests and the post-2015 development agenda as well as with multilateral forest-related agreements;

35. *Decides* that the Forum should offer to contribute to the implementation, follow-up and review of the forest-related aspects of the post-2015 development agenda, including its forest-related goals and targets;

36. *Affirms* that the Forum should also offer to contribute to the work of the high-level political forum on sustainable development;

37. *Invites* the Forum to consider, in the context of its strategic plan, its role in and contribution to the implementation of the post-2015 development agenda;

XI
Strategic Plan

38. *Decides* that the Forum should develop a concise strategic plan for the period 2017—2030 to serve as a strategic framework to enhance the coherence of and guide and focus the work of the international arrangement on forests and its components;

39. *Also decides* that the strategic plan should be aligned with the objectives of the international arrangement on forests and should incorporate a mission and vision, the global objectives on forests and the forest-related aspects of the post-2015 development agenda, taking into account significant forest-related developments in other forums, as well as identify the roles of different actors and the framework for reviewing implementation, and outline a communication strategy to raise awareness of the work of the arrangement;

40. *Requests* the Forum to operationalize the strategic plan through quadrennial programmes of work that set out priority actions and resource needs, beginning with the period 2017—2020;

XII
Review of the International Arrangement on Forests

41. *Requests* the Forum to undertake in 2024 a midterm review of the effectiveness of

the international arrangement on forests in achieving its objectives, as well as a final review in 2030, and, on that basis, to submit recommendations to the Council relating to the future course of the arrangement;

42. *Decides* that, in the context of the midterm review in 2024, the Forum could consider:

(a) A full range of options, including a legally binding instrument on all types of forests, the strengthening of the current arrangement and the continuation of the current arrangement;

(b) A full range of financing options, inter alia, the establishment of a voluntary global forest fund, in order to mobilize resources from all sources in support of the sustainable management of all types of forests;

43. *Notes* that the establishment of a global forest fund could be further considered if there is a consensus to do so at a session of the Forum prior to 2024;

XIII
Follow-up to the Eleventh Session of the Forum

44. *Decides* that the Forum should consider proposals on the following matters:

(a) Replacement of the reference to the Millennium Development Goals in paragraph 1 (b) of the non-legally binding instrument on all types of forests with an appropriate reference to the sustainable development goals and targets that will be considered by the United Nations summit for the adoption of the post-2015 development agenda, to be held in September 2015;

(b) The strategic plan for the period 2017—2030 and the quadrennial programme of work for the period 2017—2020, consistent with "Strategic Plan" of the present resolution;

45. *Invites* Member States and relevant stakeholders to provide their views and proposals on the matters referred to in paragraph 44 of the present resolution as inputs to deliberations;

46. *Decides* to establish a working group of the Forum with a time-bound mandate, for a period of up to two years in 2016 and 2017, to develop proposals on the matters referred to in paragraph 44 of the present resolution for consideration by the Forum at its special session referred to in paragraph 50 of the present resolution. The working group should:

(a) Operate in accordance with the working modalities of the Forum, as referred to in paragraph 4 of the present resolution;

(b) Elect two Co-Chairs who serve as ex officio members of the Bureau for the special session of the Forum referred in paragraph 50 of the present resolution;

47. *Also decides* that the working group of the Forum shall be convened in one session by 30 March 2017 for up to a total of five working days to develop the proposals

referred to in paragraph 44 of the present resolution;

48. *Further decides* to establish an open-ended intergovernmental ad hoc expert group to conduct up to two meetings in 2016, subject to the availability of extrabudgetary resources, to develop proposals on the matters referred to above for consideration by the working group;

49. *Requests* the Co-Chairs of the working group, under the guidance of the Bureau of the special session of the Forum, to also conduct informal consultations as needed, to facilitate a successful outcome for the working group;

50. *Decides* to hold a special session in a half-day meeting immediately upon the adjournment of the session of the working group, to consider the proposals of the working group consistent with paragraph 44 of the present resolution;

51. *Requests* the Forum to hold its next session in 2017;

XIV
Resources for the Implementation of the Present Resolution

52. *Recognizes* that the responsibilities of the secretariat of the Forum have changed considerably in their scope and complexity over the years, including in relation to servicing Forum processes and providing substantive and technical support to developing countries;

53. *Requests* the Secretary-General to continue to provide, in the most efficient and cost-effective manner, all appropriate support to the secretariat of the Forum;

54. *Urges* donor Governments and organizations, including financial institutions, and others in a position to do so, to provide voluntary contributions to the Forum Trust Fund;

55. *Calls* upon donor countries and international organizations, including financial institutions, and others in a position to do so, to provide financial support to the Forum Trust Fund in order to support the participation of developing countries, according priority to least developed countries, African States, small island developing States and countries with economies in transition, in accordance with paragraph 40 of the resolution contained in the report of the Forum on its ninth session,8 in the open-ended intergovernmental ad hoc expert group, the working group of the Forum and the sessions of the Forum;

56. *Requests* the Secretary-General to report to the Forum at its session in 2018 on the implementation of the present resolution.

1. *Rio Conventions*

"里约三公约"(*Rio Conventions*)产生于 1992 年的地球峰会(Earth Summit),包括联合国《生物多样性公约》(UNCBD)、《联合国气候变化框架公约》(UNFCCC)、《联合国防治荒漠化公约》(UNCCD)。每项公约都代表了为实现《21 世纪议程》可持续发展目标做出贡献的一种方式。这 3 项公约内在地联系在一起,在同一生态系统中运作,并解决相互依存的问题。

2. Collaborative Partnership on Forests

森林问题合作伙伴关系成立于 2001 年，支助联合国森林论坛及其成员国的工作，帮助各国加强森林的可持续管理。森林问题伙伴关系由 14 个与森林有关的国际组织组成，成员具有支助联合国森林论坛进程的实质能力、方案和资源，尤其是落实森林小组／森林论坛的行动建议。伙伴关系通过发挥各个成员的相对优势，携手支助可持续森林管理在世界各地的落实工作。

3. REDD plus

REDD+项目，全称是"Reducing Emissions from Deforestation and forest Degradation, plus the sustainable management of forests, and the conservation and enhancement of forest carbon stocks"，即"通过减少砍伐和毁坏森林而减少碳排放，加上森林的可持续管理，保护和增加森林碳储存"。这是全球减缓气候变化措施的重要组成部分。森林在减缓气候变化方面发挥的基本作用是从大气中去除二氧化碳并将其储存在生物质和土壤中。这也意味着当森林被清除或退化时，它们可以通过释放储存的碳而成为温室气体排放的来源。REDD+项目为减少因毁林和森林退化产生的温室气体排放提供补偿资金，重点是融资机制和分配机制。

Key Words and Phrases

1. accountable	/əˈkaʊntəbl/	adj.	responsible for your decision 负有责任的
2. address	/əˈdres/	v.	to deal with (something) usually skillfully or efficiently 应对
3. adjournment	/əˈdʒɜːnmənt/	n.	a temporary stopping of a trial, inquiry, or other meeting 暂停
4. align	/əˈlaɪn/	v.	to bring to or be in a state of agreement 结盟
5. catalyze	/ˈkætəlaɪz/	v.	to be the cause of (a situation, action, or state of mind) 催化；促进
6. eligible	/ˈelɪdʒəbl/	adj.	qualified 有资格的
7. fragmentation	/ˌfræɡmenˈteɪʃn/	n.	the act or process of separating into pieces usually suddenly or forcibly 分裂
8. harness	/ˈhɑːnɪs/	v.	to bring an emotion or natural source of energy under control and use it 控制
9. initiative	/ɪˈnɪʃətɪv/	v.	an important act or statement that is intended to solve a problem （重要的）法案；倡议
10. intersession	/ˌɪntəˈseʃən/	n.	a period between two academic sessions 休会期间

11. liaison	/li'eɪz(ə)n/	n.	a person who establishes and maintains communication for mutual understanding and cooperation 联络人
12. logistical	/lə'dʒɪstɪk(ə)l/	adj.	relating to the handling of the details of an operation 后勤的；组织上的
13. memorandum	/ˌmemə'rændəm/	n.	a written report that is prepared for a person or committee in order to provide them with information about a particular matter 备忘录
14. quadrennial	/kwɒd'renɪəl/	adj.	occurring or being done every four years 四年期的
15. reiterate	/ri'ɪtəreɪt/	v.	to state or do over again or repeatedly sometimes with wearying effect 重申；反复地做
16. replenishment	/rɪ'plenɪʃmənt/	n.	the process by which something is made full or complete again 补充；充满
17. subsidiary	/səb'sɪdɪəri/	adj.	related but less important or supplementary to something 辅助的
18. subsequent	/'sʌbsɪkwənt/	adj.	coming after something in time; following 随后的
19. synchronize	/'sɪŋkrənaɪz/	v.	to cause to occur or operate at the same time or rate 同步
20. synergy	/'sɪnədʒi/	n.	the interaction or cooperation of two or more organizations, substances, or other agents to produce a combined effect greater than the sum of their separate effects 协同作用

21. ad hoc expert group　　　　　　特设专家组
22. added value　　　　　　　　　附加值
23. clearing house　　　　　　　　（金融）票据交易所；信息交换所
24. ex officio member　　　　　　（因职位不需选举直接当选的）当然成员
25. joint initiative　　　　　　　　联合专门举措
26. trust fund　　　　　　　　　　信托基金

Exercises

Exercise 1 Reading Comprehension

Directions: *Read the United Nations Forum on Forests beyond 2015, and decide* whether the following statements are true or false. Write T for true or F for false in the brackets in front of each statement.

1. (   ) Dialogue, cooperation and coordination on issues related to all types of forests should be made in an integrated and holistic manner, including through cross-sectoral approaches.

2. (   ) One of the core functions of the Forum is to strengthen high-level political engagement and the participation of women and children in support of sustainable forest management in all countries.

3. (   ) To maximize the impact of the work of forum including by fostering an exchange of experiences and lessons learned among countries, regional, subregional and non-governmental partners and the Collaborative Partnership on Forests, preparatory sessional work should be enhanced.

4. (   ) At least two days' high-level segments will be held during 5 days' sessions of the Forum to accelerate action towards sustainable forest management and address forest-related global challenges and emerging issues.

5. (   ) The odd-year session focuses on the discussions on implementation and technical advice and the even-year session highlights the report of summaries of these discussions.

6. (   ) The forum assesses the progress in and makes possible proposals on strengthening the common international understanding of the concept of sustainable forest management.

Exercise 2 Skimming and Scanning

Directions: *Read the following passage excerpted from the **International Arrangement on Forests beyond 2015**. At the end of the passage, there are six statements. Each statement contains information given in one of the paragraphs of the passage. Identify the paragraph from which the information is derived. Each paragraph is marked with a letter. You may choose a paragraph more than once. Answer the questions by writing the corresponding letter in the brackets in front of each statement.*

In order to catalyze financing for implementation,

A) The Forum reiterates that there is no single solution to address all of the needs in terms of Forest financing and that a combination of actions is required at all levels, by all stakeholders and from all sources, public and private, domestic and international, bilateral and multilateral.

B) The Forum welcomes the positive work carried out by the facilitative process to date, and recognizes that it has yet to fulfil its potential as set out in the resolutions contained in the reports of the Forum on the special session of its ninth session and on its ninth session;

C) The facilitative process should promote the design of national forest financing strategies to mobilize resources for sustainable forest management, including existing national initiatives, within the framework of national forest programmes or their equivalent, to facilitate access to existing and emerging financing mechanisms, including the Global Environment Facility and the Green Climate Fund, consistent with their mandates, in order to implement sustainable forest management;

D) The facilitative process should serve as a clearing house on existing, new and emerging financing opportunities and as a tool for sharing lessons learned from successful projects, building on the Collaborative Partnership on Forests online sourcebook for forest financing;

E) The facilitative process should ensure that special consideration is given to the special needs and circumstances of Africa, the least developed countries, low-forest-cover countries, high-forest-cover countries, medium-forest-cover low-deforestation countries and small island developing States, as well as countries with economies in transition, in gaining access to funds;

F) The Forum also decides to request the secretariat, in consultation with the members of the Forum and the Collaborative Partnership on Forests, to make recommendations on ways to further increase the effectiveness and efficiency of the operation of the strengthened facilitative process and submit them for consideration by the Forum at its session in 2018;

G) The Forum decides to welcome the report of the secretariat of the Global Environment Facility to the Forum on the mobilization of financial resources through the sustainable forest management/REDD-plus incentive programme under the fifth replenishment of the Facility, and invite the secretariat of the Facility to periodically provide information on the mobilization of financial resources and funds that are dedicated to sustainable forest management;

H) The Forum decides to also welcome the decision taken by the Assembly of the Global Environment Facility at its session in May 2014 to include a sustainable forest management strategy in the sixth replenishment period of the Facility (2014—2018) to support the sustainable management of all types of forests;

I) he Forum decides to encourage eligible Member States, taking into account the cross-sectoral nature of sustainable forest management, to make full use of the existing potential of the sustainable forest management strategy under the sixth replenishment of the Global Environment Facility to harness synergies across the focal areas of the Facility in order to further reinforce the importance of sustainable forest management for integrating environmental and development aspirations;

1. (　　) Eligible Member States are encouraged to make full use of the existing potential to reinforce the importance of sustainable management for integrating environmental and development aspirations.

2. (　　) The facilitative process is responsible for offering existing, new and emerging financing opportunities and sharing lessons learned from successful projects.

3. (　　) The Forum invites the secretariat of the Facility to periodically provide information on the mobilization of financial resources and funds.

4. (　　) Stakeholders and all sources should cooperate to meet the needs in terms of forest financing.

5. (　　) The least developed countries and countries with economies in transition should be given special consideration in gaining access to funds.

6. (　　) The Facilitative process still has much work to do in order to fulfil its potential set out in the resolutions contained in the reports of the Forum.

Exercise 3　Word Formation

Directions: *In this section, there are ten sentences from the **International Arrangement on Forests beyond 2015**. You are required to complete these sentences with the proper form of the words given in the blanks.*

1. The Economic and Social Council acknowledged the progress made by countries and stakeholders towards ＿＿＿＿＿＿＿＿＿＿ forest management. (sustain)

2. UNFF intends to ＿＿＿＿＿＿＿＿＿＿ long-term political commitment to the achievement of the objectives listed in the present resolution. (strength)

3. One of the core functions of the ＿＿＿＿＿＿＿＿＿＿ Partnership on Forests is to promote the implementation of the non-legally binding instrument on all types of forests. (*collaborate*)

4. The Forum should restructure its sessions and enhance its intersessional work to ＿＿＿＿＿＿＿＿＿＿ the impact and relevance of its work. (maximum)

5. The specific task is to assess the progress in, and make possible proposals on, the ＿＿＿＿＿＿＿＿＿＿ of the resolutions and decisions of the Forum. (implement)

6. UNFF reaffirms that there is no single ＿＿＿＿＿＿＿＿＿＿ to address all of the needs in terms of forest financing. (solve)

7. The capacity of the secretariat should be enhanced to ＿＿＿＿＿＿＿＿＿＿ and effectively manage the strengthened facilitative process. (efficiency)

8. The Forum shall invite the secretariat of the facility to periodically provide information on the ＿＿＿＿＿＿＿＿＿＿ of financial resources and funds. (mobilize)

9. The Forum should promote coherence, ＿＿＿＿＿＿＿＿＿＿ and cooperation on forest-related issues. (coordinate)

10. All member organizations are encouraged to ＿＿＿＿＿＿＿＿＿＿ ways to actively involve major groups in activities of the Partnership. (identity)

Exercise 4　Translation

Section A

Directions: *Read the **International Arrangement on Forests beyond 2015**, and complete the sentences by translating into English the Chinese given in the blanks.*

1. Assess the progress in enhancing policy coherence, dialogue and cooperation on forests, fostering synergies in global forest-related processes and strengthening the common international understanding of ＿＿＿＿＿＿＿＿＿＿ (可持续森林经营概念) as set out in the non-legally binding instrument on all types of forests.

2. Establish ＿＿＿＿＿＿＿＿＿＿ (不限制成员名额的政府间特设专家组) to conduct up to two meetings in 2016, subject to the availability of extrabudgetary resources, to develop proposals on the matters referred to above for consideration by the

working group.

3. It should promote the design of national forest financing strategies to mobilize resources for sustainable forest management, to facilitate access to _____ (现有和新出现的筹资机制) in order to implement sustainable forest management.

4. Ensure that special consideration is given to the special needs and circumstances of Africa, the least developed countries, medium-forest-cover low-deforestation countries and small island developing States, as well as _____ (经济转型国家) in gaining access to funds.

5. Invite Member States to continue to monitor and assess progress towards implementing sustainable forest management, including the non-legally binding instrument on all types of forests and the global objectives on forests, and _____ (在自愿基础上提交国家进展报告) to the Forum.

6. Manage _____ (全球森林筹资促进网络) and implement its activities in collaboration with relevant members of the Collaborative Partnership on Forests.

7. Strengthen the Partnership by formalizing its working modalities, including through consideration of _____ (一项多边谅解备忘录) and by developing procedures for its effective functioning and operation.

8. Invite the governing bodies of member organizations to _____ (在工作方案中列入专项资金) to support Partnership activities, as well as budgeted activities supporting the priorities of the Forum as outlined in the strategic plan referred to in "Strategic Plan" of the present resolution, consistent with their mandates.

9. The strategic plan should incorporate a mission and vision, the global objectives on forests, taking into account significant forest-related developments in other forums, as well as identify the roles of different actors and the framework for reviewing implementation, and _____ (制定宣传策略以提高认识) of the work of the arrangement.

10. Forum should offer to contribute to _____ (实施、跟进、审查与森林问题有关的方面) of the post-2015 development agenda, including its forest-related goals and targets.

Section B

Directions: *Translate the following sentences from English into Chinese.*

1. Calls upon Member States, as well as other members of the governing bodies of member organizations of the Collaborative Partnership on Forests, to support the work of the Partnership, including by considering dedicated funding for Partnership activities consistent with the respective mandates of Partnership members as an essential strategy for improving cooperation, synergies and coherence on forest issues at all levels. (*Collaborative Partnership*)

2. Recognizing the achievements of the international arrangement on forests since its inception, in particular the adoption by the General Assembly of the non-legally binding

instrument on all types of forests adopted by the Forum, as well as the contributions of the Collaborative Partnership on Forests, while stressing the continued challenges of and the need to strengthen the international arrangement on forests and to continue efforts to contribute to the promotion and implementation of sustainable forest management. (*Preamble*)

3. The international arrangement on forests beyond 2015 should operate in a transparent, effective, efficient and accountable manner and should provide added value and contribute to enhanced coherence, cooperation and synergies with respect to other forest-related agreements, processes and initiatives. (*International Arrangemeat on Forests beyond 2015*)

4. The core function of the Forum is to provide a coherent, open, transparent and participatory global platform for policy development, dialogue, cooperation and coordination on issues related to all types of forests, including emerging issues, in an integrated and holistic manner, including through cross-sectoral approaches. (*United Nations Forum on Forests beyond 2015*)

5. Encourage eligible Member States, taking into account the cross-sectoral nature of sustainable forest management, to make full use of the existing potential of the sustainable forest management strategy under the sixth replenishment of the Global Environment Facility to harness synergies across the focal areas of the Facility in order to further reinforce the importance of sustainable forest management for integrating environmental and development aspirations. (*Catelysing Financing for Implementation*)

6. Calls upon donor countries and international organizations, including financial institutions, and others in a position to do so, to provide financial support to the Forum Trust Fund in order to support the participation of developing countries, according priority to least developed countries, African States, small island developing States and countries with economies in transition, in accordance with paragraph 40 of the resolution contained in the report of the Forum on its ninth session, in the open-ended intergovernmental ad hoc expert group, the working group of the Forum and the sessions of the Forum. (*Resourcos for the Implementation of the Present Resolution*)

Extensive Readings

Passage 1

Directions: *Read the following passage and choose the best answer for each of the following questions according to the information given in the passage.*

Forests are nature's most bountiful and versatile renewable resource, providing simultaneously a wide range of economic, social, environmental and cultural benefits and services. The worldwide demand for their numerous functions and outputs is increasing with the expanding population, while the global forest resource is shrinking either as a result of overharvesting, deforestation and permanent conversion to other forms of land use in many tropical regions, or as a consequence of forest decline associated with airborne pollutants in temperate regions.

Forests represent a unique situation in terms of global environmental issues. Physically, they are located within the territories of sovereign states, yet their environmental role extends beyond their borders at both transboundary and regional as well as global levels. For example, the management, or mismanagement, of watershed forests of international rivers has transboundary implications in terms of soil and water conservation in neighboring countries. Similarly, airborne pollutants generated in one country may be transported across the boundary and cause forest decline in others. The role of forests in global ecological cycles highlights the environmental significance of forests beyond the boundaries of the nations where they are located. In this context, they are being viewed as global commons similar to the atmosphere and oceans.

Conservation and sustainable development of all types of forests worldwide have now emerged as priority items on the international policy agenda, particularly in the context of the United Nations Conference on Environment and Development (UNCED), to be held in Brazil in June 1992. The role of forests is receiving particular attention in the biodiversity and climate change conventions currently under negotiation. While special interest groups are only focusing on a specific role or function of forests (e.g. as a reservoir of biodiversity, for carbon sequestration, economic development, subsistence, fuel, etc.), national and international policy-makers face the challenge of reconciling the role of forests in meeting national socioeconomic and environmental objectives as well as the global environmental and socio-economic interests of the community of nations. Ecological considerations are now being viewed not as subordinate but as an integral part of economic policy and planning.

Sustainable forest development is also emerging as a consideration in the international trade of forest products. Many consumers, individually and collectively, are preferring to buy products obtained from sustainably managed forests and manufactured by environmentally acceptable processes. There have been consumer threats to boycott wood products that are not "green" both in terms of raw materials and manufacturing processes.

In contrast, most members of the forestry community have usually dealt with local issues and with "delivering wood to the mill gate". The national and international forestry community is relatively inexperienced, both technically and politically, in dealing with the globalization of forest-related issues. Consequently, their participation in these deliberations and their influence in shaping the international forestry agenda to date have been marginal. Forestry, involving long-term commitments, usually receives limited political attention in comparison with most other, often shorter-term, socio-economic policies. The current attention being paid to forest related issues by international political communities should be viewed as a rare window of opportunity to advance the interests for forestry of political support and sustainable forest development, and to promote the multiple benefits provided by forests. These benefits range from meeting the socio-economic needs of forest dwellers, forest-based communities and forest industry to conserving environmental values.

It is important to understand the evolution of the structure and content of international deliberations on forests, the shifts in our values and the consequent impact on forestry practices. The forestry and scientific communities are faced with the challenge of defining sustainable forest development, formulating a conceptual framework and establishing internationally accepted criteria and approaches for the practice of sustainable forest development to meet multiple human needs.

The term "environmentally sustainable economic development", more commonly known as "sustainable development", has been popularized globally by the report of the World Commission on Environment and Development (WCED), *Our Common Future*. In this report, sustainable development is defined as "economic development that meets the needs of the present without compromising the ability of future generations to meet their own needs". The term sustainable development has captured the imagination both of the public and politicians at local, national, regional and international levels, and has instigated much discussion. However, there have been limited attempts to put the concept into practice.

The forestry sector, perhaps more than any other, is well positioned to provide worldwide leadership in the practice of sustainable development. The forestry community is accustomed to a long-term perspective; it is reasonably knowledgeable about the response of forest ecosystems to natural and human disturbances; it is comfortable with the sustained yield principle; and, in a few instances, it has attempted to practice a multiple and integrated use of forests. As compared to many other industrial sectors, it is relatively easier for the forest community to expand its scope from sustained yield to sustainable development, which requires a shift from forest management to forest ecosystem management. Sustainable development of forest land and its multiple economic and environmental values involves maintaining indefinitely, without unacceptable impairment, the productive and renewal capacities as well as the species and ecological diversity of forest ecosystems.

1. What is the factor leading to the shrinking of global forest resource? _____
    A. Overexploitation.　　　　　　　　B. Afforestation.
    C. Land degradation.　　　　　　　　D. Water pollution.
2. Why are forests viewed as global commons? _____
    A. Because trees are located in the territories of sovereign states in the world.
    B. Because the ways of the management of watershed forests is the same.
    C. Because the environmental role of forests is transboundary.
    D. Because the effect of pollution on forest exists in all countries.
3. What is listed as the priority item on international policy agenda? _____
    A. Social and economic development.　　B. Sustainable development of trees.
    C. Carbon sequestration and stock.　　　D. Climate change in the world.
4. The following statements about forestry community are **TRUE** except _____.
    A. forestry community is inexperienced technically and politically
    B. forestry community's influence in shaping forestry agenda is little

C. forestry community is facing a good opportunity to get political support
D. forestry community usually receives more political attention

5. Which is **NOT** the challenge that the forestry and scientific communities are facing? ___
   A. How to shift from forest management to ecosystem management.
   B. How to formulate a conceptual framework.
   C. How to establish internationally accepted criteria.
   D. How to define sustainable forest development.

Passage 2

Directions: *In this section, there is a passage with twelve blanks. You are required to select one word for each blank from a list of choices given in a word bank following the passage. Read the passage through carefully before making your choices. Each choice in the bank is identified by a letter. You may not use any of the words in the bank more than once.*

A. indigenous	B. sustainably	C. mitigate	D. innovative
E. awareness	F. ecosystems	G. partnerships	H. disasters
I. contribute	J. addressing	K. achieve	L. communities

Sustainable Development Goal 15 aims to protect, restore and promote sustainable use of terrestrial ecosystems, sustainably manage forests, combat desertification, and halt and reverse land degradation and halt biodiversity loss.

Forests have a significant role in reducing the risk of natural ___1___, including floods, droughts, landslides and other extreme events. At global level, forests ___2___ climate change through carbon sequestration, ___3___ to the balance of oxygen, carbon dioxide and humidity in the air and protect watersheds, which supply 75% of freshwater worldwide.

Investing in forests and forestry represent an investment in people and their livelihoods, especially the rural poor, youth and women. Around 1.6 billion people—including more than 2,000 ___4___ cultures—depend on forests for their livelihood. Forests are the most biologically-diverse ___5___ on land, home to more than 80% of the terrestrial species of animals, plants and insects. They also provide shelter, jobs and security for forest-dependent ___6___.

Therefore, the future of forests and forestry in sustainable development at all levels was at the core of the XIV World Forestry, hosted in Durban from 7 to 11 September 2015. The Durban Declaration called for new ___7___ among forest, agriculture, finance, energy, water and other sectors, as well the engagement with indigenous people and local community.

The importance of investing in world's forests and of taking "political commitment at the highest levels, smart policies, effective law enforcement, ___8___ partnerships and funding" was also recalled by the former UN Secretary-General Mr Ban Ki-moon in his Message on the occasion of the 2015 International Day of Forests.

Both the International Day of Forests and the International Year of Forest aimed at raising ___9___ on the importance of all types of forests and of trees outside forests.

Prior to the *2030 Agenda* for Sustainable Development, the outcome document of the Rio+20 Conference, *The Future We Want*, stress the importance of improving the livelihoods of people and communities by creating the conditions required to ____10____ manage forests. It also recognizes the role of the UN Forum on Forests in ____11____ forest-related issues in a holistic and integrated manner, and in promoting international policy coordination and cooperation in order to ____12____ forest management. It calls for the mainstreaming of sustainable forest management and practices into economic policy and decision-making.

Further Studies and Post-Reading Discussion

Task 1
Directions: *Surf the Internet and find more information about the* **International Arrangement on Forests beyond 2015**. *Work in groups and work out a report on one of the following topics:*
1. Odd-year and even-year sessions of the Forum.
2. China's achievements in implementing the *United Nations* Forest Instrument.

Task 2
Directions: *Read the following sentences on Eco-Civilization and make a speech on your understanding of the eco-environmental conservation.*

Global Environmental Governance

2021年4月，习近平在领导人气候峰会上全面系统阐释推进全球生态环境治理、构建人与自然生命共同体的核心要义，即坚持人与自然和谐共生，坚持绿色发展，坚持系统治理，坚持以人为本，坚持多边主义，坚持共同但有区别的责任原则。这"六个坚持"主张，是对人类经济社会实践经验的提炼升华，为处于关键节点的全球生态环境治理贡献了中国智慧。

从推动达成气候变化《巴黎协定》到全面履行《联合国气候变化框架公约》，从大力推进绿色"一带一路"建设到深度参与全球生态环境治理，中国作为全球生态文

At the Leaders Summit on Climate in April 2021, Xi Jinping shared his thoughts on improving global environmental governance and fostering a community of life for humanity and nature. He called on the international community to commit to harmony between humanity and nature, green development, systemic governance, a people-centered approach, multilateralism, and the principle of common but differentiated responsibilities. This is a summary of human experience in economic and social development that will work well in promoting global environmental governance.

From the adoption of the Paris Agreement to its own implementation of the UN Framework Convention on Climate Change, from green cooperation among Belt and Road countries to its own parti-

明建设的参与者、贡献者、引领者，一直为建设清洁美丽的世界砥砺前行。（摘自《中国关键词》生态文明篇）	cipation in global environmental governance, China has always been a participant in, contributor to and leader of the global endeavor to build an eco-friendly civilization and a beautiful world. (Excerpt from *keywords to Understand China on Eco-Civilization*)

第 8 章　联合国森林战略规划
（2017—2030 年）

Chapter 8　United Nations Strategic Plan for Forests 2017—2030

Background and Significance

《联合国森林战略规划（2017—2030 年）》于 2017 年 4 月 27 日在第 71 届联合国大会审议通过，这是首次以联合国名义做出的全球森林发展战略。规划阐述了 2030 年全球林业发展愿景与使命，制定了全球森林目标和行动领域，提出了各层级开展行动的执行框架和资金手段，明确了实现全球森林目标的监测、评估和报告体系，并提出五大专项行动：在地方、国家、区域和国际层面加大力度，支持可持续利用和保护森林，包括投资于宣传教育行动，以提高公众对森林重要性的认识，帮助人们改变破坏性行为；必须确保将可持续的森林和土地管理纳入国家发展规划和预算进程之中；加强现有伙伴关系并建立新的创新合作机制，将政府、国际组织、民间社会、土地所有者、私营部门、地方社区以及环境、科学和学术机构团结起来，共同制订促进可持续经济发展和环境保护的有效政策计划；作为全面保护森林战略的一部分，帮助森林依赖型社区扩大不基于森林的经济和社会发展机会，并为其提供支持生计的替代来源；积极寻求利用科学、创新和技术的力量来推动解决毁林的根源性问题。《联合国森林战略规划（2017—2030 年）》对全球森林发展具有重要的指导意义和导向作用。

Text Study

United Nations Strategic Plan for Forests 2017—2030

I. Introduction

A. Vision and mission

1. Forests are among the world's most productive land-based ecosystems and are

essential to life on earth. The United Nations strategic plan for forests 2017—2030 provides a global framework for action at all levels to sustainably manage all types of forests and trees outside forests, and to halt deforestation and forest degradation. The strategic plan also provides a framework for forest-related contributions to the implementation of *the 2030 Agenda for Sustainable Development*[①], *the Paris Agreement*[②] adopted under the *United Nations Framework Convention on Climate Change*[③], the *Convention on Biological Diversity*[④], the *United Nations Convention to Combat Desertification in Those Countries Experiencing Serious Drought and/or Desertification, Particularly in Africa*[⑤], the *United Nations Forest Instrument*[⑥] and other international forest-related instruments, processes, commitments and goals.

2. The strategic plan serves as a reference framework for the forest-related work of the United Nations system and for the fostering of enhanced coherence, collaboration and synergies among United Nations bodies and partners towards the vision and mission set out below. It also serves as a framework to enhance the coherence of and guide and focus the work of the international arrangement on forests and its components.

Shared United Nations vision

3. The shared United Nations vision is of world in which all types of forests and trees outside forests are sustainably managed, contribute to sustainable development and provide economic, social, environmental and cultural benefits for present and future generations.

Shared United Nations mission

4. The shared United Nations mission is to promote sustainable forest management and the contribution of forests and trees outside forests to *the 2030 Agenda for Sustainable Development*, including by strengthening cooperation, coordination, coherence, synergies and political commitment and action at all levels.

B. Importance of forests to people and *the 2030 Agenda for Sustainable Development*

5. Forests cover 30% of the Earth's land area, or nearly 4 billion hectares, and are essential to human well-being, sustainable development and the health of the planet.[⑦] An estimated 1.6 billion people, or 25% of the global population, depend on forests for

---

① General Assembly resolution 70/1.

② See FCCC/CP/2015/10/Add.1, decision 1/CP.21, annex.

③ United Nations, *Treaty Series*, vol. 1771, No. 30822.480.

④ Ibid., vol. 1760, No. 30619.

⑤ Ibid., vol. 1954, No. 33.

⑥ *The United Nations Forest Instrument* was adopted by the United Nations Forum on Forests and the General Assembly in 2007. It sets out four shared global objectives on forests and 44 national and international policies, measures and actions to implement sustainable forest management and enhance the contribution of forests to *the 2030 Agenda for Sustainable Development*.

⑦ For a glossary of forest-related definitions, see the terms and definitions for the most recent *Global Forest Resources Assessment* of FAO.

subsistence, livelihood, employment and income generation.

6. Forests provide essential ecosystem services, such as timber, food, fuel, fodder, non-wood products and shelter, as well as contribute to soil and water conservation and clean air. Forests prevent land degradation and desertification and reduce the risk of floods, landslides, avalanches, droughts, dust-storms, sandstorms and other natural disasters. Forests are home to an estimated 80% of all terrestrial species. Forests contribute substantially to climate change mitigation and adaptation and to the conservation of biodiversity.

7. When sustainably managed, all types of forests are healthy, productive, resilient and renewable ecosystems, providing essential goods and services to people worldwide. In many regions, forests also have important cultural and spiritual value. As set out in *the United Nations Forest Instrument*, "Sustainable forest management, as a dynamic and evolving concept, is intended to maintain and enhance the economic, social and environmental values of all types of forests, for the benefit of present and future generations".

8. The sustainable management of forests and trees outside forests is vital to the integrated implementation of *the 2030 Agenda for Sustainable Development*, including the achievement of the Sustainable Development Goals, especially Goal 15 ("Sustainably manage forests, combat desertification, halt and reverse land degradation, halt biodiversity loss").

9. In recognition of the extraordinary importance of forests to people, the General Assembly, in its resolution 67/200, proclaimed 21 March as the International Day of Forests, which is celebrated around the world each year to raise awareness and promote action on forest issues.

C. Trends and challenges

10. Despite the crucial contribution of forests to life on earth and human well-being, deforestation and forest degradation continue in many regions, often in response to the demand for wood, food, fuel and fibre. Many drivers of deforestation lie outside the forest sector and are rooted in wider social and economic issues, including challenges related to reducing poverty, urban development and policies that favour land uses that produce higher and more rapid financial returns, such as agriculture, energy, mining and transportation.

11. Forests are also at risk from illegal or unsustainable logging, unmanaged fires, pollution, dust-storms, sandstorms and wind-storms, disease, pests, invasive alien species, fragmentation and the impact of climate change, including severe weather events, all of which threaten the health of forests and their ability to function as productive and resilient ecosystems.

12. Continued rapid population growth, as well as rising per capita income, is accelerating the global demand for and consumption of forest products and services and putting pressure on forests. With the world population projected to reach 6 billion by 2050, meeting future demand for forest products and services depends on urgent action and cross-sectoral policy coordination at all levels to secure sustainable forest management, including forest conservation, restoration and expansion.

13. At the global level, there is a need to reduce fragmentation and enhance coordination

among the many international organizations, institutions and instruments addressing forest issues.

14. At the national, local and regional levels, cross-sectoral coordination on forests can be weak, and forest authorities and stakeholders may not be full partners in land use planning and development decisions.

15. The effective implementation of sustainable forest management is critically dependent upon adequate resources, including financing, capacity development and the transfer of environmentally sound technologies and, in particular, the need to mobilize increased financial resources, including from innovative sources, for developing countries, including least developed countries, landlocked developing countries and small island developing States, as well as countries with economies in transition. Implementation of sustainable forest management is also critically dependent upon good governance at all levels.

D. Opportunities for enhanced and value-added action on sustainable forest management

16. The launch of the United Nations strategic plan for forests 2017—2030 comes at a time of unprecedented opportunity for strengthened and decisive action by all actors at all levels, within and beyond the United Nations system, to safeguard the world's forests and their multiple values, functions and benefits, now and in the future.

17. The strategic plan is aimed at building on the momentum provided by the 2015 global milestones represented by the adoption of *the 2030 Agenda for Sustainable Development*, the *Addis Ababa Action Agenda*[①] and *the Paris Agreement* under *the United Nations Framework Convention on Climate Change*.

18. The United Nations system can contribute to these initiatives and achieve the vision and mission for forests by advancing a set of global goals and targets in support of the sustainable management of all types of forests and trees outside forests.

E. International arrangement on forests

19. The international arrangement on forests is composed of the United Nations Forum on Forests, a functional commission of the Economic and Social Council, and the 197 States members thereof, the secretariat of the Forum, the Collaborative Partnership on Forests, the Global Forest Financing Facilitation Network and the Trust Fund for the United Nations Forum on Forests. The Forum is the United Nations body mandated to address forest-related issues in an integrated and holistic manner and oversees implementation of the strategic plan and the United Nations forest instrument, as well as operation of the Global Forest Financing Facilitation Network.

20. The work of the Forum is supported by its secretariat, the Trust Fund for the United Nations Forum on Forests and the Collaborative Partnership on Forests. The Collaborative Partnership on Forests is a voluntary partnership chaired by the Food and Agriculture Organization of the United Nations (FAO) and comprised of 14 international organizations

---

① See General Assembly resolution 69/313, annex.

with significant programmes on forests①. The functions of the Forum, its secretariat and the Collaborative Partnership are contained in Economic and Social Council resolution 2015/33.

21. The international arrangement on forests involves as partners a variety of international, regional, subregional and non-governmental organizations and processes with forest-related programmes, and recognizes the important role of major groups and other relevant stakeholders at all levels in promoting and achieving sustainable forest management worldwide.

22. The objectives of the international arrangement on forests are to:

(a) Promote the implementation of sustainable management of all types of forests, in particular the implementation of *the United Nations Forest Instrument*;

(b) Enhance the contribution of all types of forests and trees outside forests to *the 2030 Agenda for Sustainable Development*;

(c) Enhance cooperation, coordination, coherence and synergies on forest-related issues at all levels;

(d) Foster international cooperation, including North-South, South-South, North-North and triangular cooperation, as well as public-private partnerships and cross-sectoral cooperation, at all levels;

(e) Support efforts to strengthen forest governance frameworks and means of implementation, in accordance with *the United Nations Forest Instrument*, in order to achieve sustainable forest management.

Ⅱ. Global Forest Goals and Targets

23. At the heart of the strategic plan are six global forest goals and 26 associated targets to be achieved by 2030. These goals and targets, set out below, fully encompass and build on the solid foundation provided by the four global objectives on forests included in *the United Nations Forest Instrument*.

24. The global forest goals and targets are voluntary and universal. They support the objectives of the international arrangement on forests and are aimed at contributing to progress on the Sustainable Development Goals, the Aichi Biodiversity Targets②, *the Paris Agreement* adopted under the *United Nations Framework Convention on Climate Change* and other international forest-related instruments, processes, commitments and goals.

25. The vision, principles and commitments set out in *the 2030 Agenda for Sustainable Development* provide the context for the global forest goals and targets, which are interconnected and integrate the economic, social and environmental dimensions of sustainable forest management and sustainable development.

26. The global forest goals and targets are intended to stimulate and provide a framework

---

① For a list of member organizations of the Collaborative Partnership on Forests, see the website of the Partnership.

② United Nations Environment Programme, document NEP/CBD/COP/10/27, annex, decisionⅩ/2, annex.

for voluntary actions, contributions and enhanced cooperation by countries and international, regional, subregional and non-governmental partners and stakeholders. They also provide a reference for enhanced coherence and collaboration on forests within the United Nations system and among member organizations of the Collaborative Partnership on Forests, as well as among other forest-related organizations and processes.

27. The global goals and targets encompass a wide variety of thematic areas in regard to which voluntary actions, contributions and cooperation are needed to advance their achievement. These thematic areas reflect and encompass the 44 policies, measures and actions set out in the United Nations forest instrument. A non-exhaustive list of indicative thematic areas for action is contained in the appendix to the present document. Indicative thematic areas may correspond to more than one goal.

Global forest goal 1

Reverse the loss of forest cover worldwide through sustainable forest management, including protection, restoration, afforestation and reforestation, and increase efforts to prevent forest degradation and contribute to the global effort of addressing climate change.

1.1 Forest area is increased by 3% worldwide[①].

1.2 The world's forest carbon stocks are maintained or enhanced.

1.3 By 2020, promote the implementation of sustainable management of all types of forests, halt deforestation, restore degraded forests and substantially increase afforestation and reforestation globally.

1.4 The resilience and adaptive capacity of all types of forests to natural disasters and the impact of climate change is significantly strengthened worldwide.

Goal 1 and its targets support and contribute to the achievement of, among other things, Sustainable Development Goal targets 6.6, 12.2, 13.1, 13.3, 14.2, 15.1–15.4 and 15.8, as well as Aichi Biodiversity targets 5, 7, 9, 11, 14 and 15.

Global forest goal 2

Enhance forest-based economic, social and environmental benefits, including by improving the livelihoods of forest-dependent people.

2.1 Extreme poverty for all forest dependent people is eradicated.

2.2 Increase the access of small-scale forest enterprises, in particular in developing countries, to financial services, including affordable credit, and their integration into value chains and markets.

2.3 The contribution of forests and trees to food security is significantly increased.

2.4 The contribution of forest industry, other forest-based enterprises and forest ecosystem services to social, economic and environmental development, among other things, is significantly increased.

2.5 The contribution of all types of forests to biodiversity conservation and climate change mitigation and adaptation is enhanced, taking into account the mandates and ongoing

---

① Based on the *Global Forest Resources Assessment 2015*.

work of relevant conventions and instruments.

Goal 2 and its targets support and contribute to the achievement of, among other things, Sustainable Development Goal targets 1.1, 1.4, 2.4, 4.4, 5.a, 6.6, 8.3, 9.3, 12.2, 12.5, 15.6 and 15.c, as well as Aichi Biodiversity targets 4, 14 and 18.

Global forest goal 3

Increase significantly the area of protected forests worldwide and other areas of sustainably managed forests, as well as the proportion of forest products from sustainably managed forests.

3.1 The area of forests worldwide designated as protected areas or conserved through other effective area-based conservation measures is significantly increased.

3.2 The area of forests under long-term forest management plans is significantly increased.

3.3 The proportion of forest products from sustainably managed forests is significantly increased.

Goal 3 and its targets support and contribute to the achievement of, among other things, Sustainable Development Goal targets 7.2, 12.2, 12.6, 12.7, 14.2, 14.5, 15.2 and 15.4, as well as Aichi Biodiversity targets 7, 11, 12 and 16.

Global forest goal 4

Mobilize significantly increased, new and additional financial resources from all sources for the implementation of sustainable forest management and strengthen scientific and technical cooperation and partnerships.

4.1 Mobilize significant resources from all sources and at all levels to finance sustainable forest management and provide adequate incentives to developing countries to advance such management, including for conservation and reforestation.

4.2 Forest-related financing from all sources at all levels, including public (national, bilateral, multilateral and triangular), private and philanthropic financing, is significantly increased.

4.3 North-South, South-South, North-North and triangular cooperation and public- rivate partnerships on science, technology and innovation in the forest sector are sig- nificantly enhanced and increased.

4.4 The number of countries that have developed and implemented forest financing strategies and have access to financing from all sources is significantly increased.

4.5 The collection, availability and accessibility of forest-related information is improved through, for example, multidisciplinary scientific assessments.

Goal 4 and its targets support and contribute to the achievement of, among other things, Sustainable Development Goal targets 12.a, 15.7, 15.a, 15.b, 17.1–17.3, 17.6–17.7 and 17.16–17.19, as well as Aichi Biodiversity target 19.

Global forest goal 5

Promote governance frameworks to implement sustainable forest management, including through *the United Nations Forest Instrument*, and enhance the contribution of forests to *the*

2030 Agenda for Sustainable Development.

5.1 Number of countries that have integrated forests into their national sustainable development plans and/or poverty reduction strategies is significantly increased.

5.2 Forest law enforcement and governance are enhanced, including through significantly strengthening national and subnational forest authorities, and illegal logging and associated trade are significantly reduced worldwide.

5.3 National and subnational forest-related policies and programmes are coherent, coordinated and complementary across ministries, departments and authorities, consistent with national laws, and engage relevant stakeholders, local communities and indigenous peoples, fully recognizing the *United Nations Declaration on the Rights of Indigenous Peoples*[①].

5.4 Forest-related issues and the forest sector are fully integrated into decision- making processes concerning land use planning and development.

Goal 5 and its targets support and contribute to the achievement, among other things, of Sustainable Development Goal targets 1.4, 2.4, 5.a, 15.9, 15.c, 16.3, 16.5-16.7, 16.10 and 17.14, as well as Aichi Biodiversity targets 2 and 3.

Global forest goal 6

Enhance cooperation, coordination, coherence and synergies on forest-related issues at all levels, including within the United Nations system and across member organizations of the Collaborative Partnership on Forests, as well as across sectors and relevant stakeholders.

6.1 Forest-related programmes within the United Nations system are coherent and complementary and integrate the global forest goals and targets, where appropriate.

6.2 Forest-related programmes across member organizations of the Collaborative Partnership on Forests are coherent and complementary and together encompass the multiple contributions of forests and the forest sector to *the 2030 Agenda for Sustainable Development.*

6.3 Cross-sectoral coordination and cooperation to promote sustainable forest management and halt deforestation and forest degradation are significantly enhanced at all levels

6.4 A greater common understanding of the concept of sustainable forest management is achieved and an associated set of indicators is identified.

6.5 The input and involvement of major groups and other relevant stakeholders in the implementation of the strategic plan and in the work of the Forum, including intersessional work, is strengthened.

Goal 6 and its targets support and contribute to the achievement of, among other things, Sustainable Development Goal target 17.14.

Ⅲ. Implementation Framework

28. The United Nations strategic plan for forests 2017—2030 provides a reference for

---

① General Assembly resolution 61/295, annex.

ambitious and transformational actions by all actors, at all levels, to achieve its global forest goals and targets. An overview of roles and responsibilities and means of implementation is outlined below.

A. Roles and responsibilities

Members of the Forum

29. The individual and collective actions and commitments of members of the Forum are decisive for the successful implementation of the strategic plan and achievement of its global forest goals and targets.

30. Members may, on a voluntary basis, determine their contributions towards achieving the global forest goals and targets, taking into account their national circumstances, policies, priorities, capacities, levels of development and forest conditions.

31. Members may include in their voluntary national contributions, as appropriate, the forest-related contributions they intend to make with regard to other international forest-related commitments and goals, such as the implementation of the *2030 Agenda* and its Sustainable Development Goals, the *Aichi Biodiversity Targets* and actions to address climate change under *the Paris Agreement* adopted under the *United Nations Framework Convention on Climate Change*.

32. Members may, on a voluntary basis, communicate their progress on the voluntary national contributions to the United Nations Forum on Forests at regular intervals determined by the Forum, in accordance with paragraph 67 of the strategic plan for forests. In order to avoid any additional reporting burden, such voluntary communications on their voluntary national contributions may be part of their voluntary reporting on the strategic plan and *the United Nations Forest Instrument*.

33. Members of the Forum, as members of the governing bodies of forest-related international, regional and subregional organizations and processes, as appropriate, are encouraged to promote the integration of the global forest goals and targets into the strategies and programmes of these organizations, processes and instruments, consistent with their mandates and priorities.

United Nations Forum on Forests and its secretariat

34. As part of the United Nations system and the international arrangement on forests, the Forum, in carrying out its core functions as defined in Economic and Social Council resolution 2015/33, should be guided by the strategic plan for forests. The Forum's quadrennial programmes of work are to reflect its contribution to the global forest goals and targets for each quadrennium.

35. The Forum is the responsible intergovernmental body for follow up and review of the implementation of the strategic plan, including through providing guidance to the Collaborative Partnership on Forests and ensuring the smooth interplay between its odd- and even-year sessions.

36. The Forum secretariat services and supports the Forum in all matters related to the Forum's quadrennial programmes of work and the strategic plan.

37. The Forum should structure its annual sessions and enhance its intersessional activities to maximize the impact and relevance of its work under the quadrennial programmes of work, including by fostering cross-sectoral exchanges to enhance synergies inside and outside the United Nations.

Collaborative Partnership on Forests and its member organizations

38. Member organizations of the Collaborative Partnership on Forests play an important role in implementing the strategic plan and are encouraged to integrate relevant global forest goals and targets into their forest-related plans and programmes, where appropriate and consistent with their respective mandates.

39. The Partnership is invited to support the Forum and its members in advancing the global forest goals and targets, including through cooperation and partnership among its members, implementing a joint workplan with the Partnership which is aligned with the Forum's quadrennial programmes of work and identifying collective actions by all or subsets of the Partnership's members, as well as associated resource needs.

40. Members of the Forum are encouraged to support the Partnership workplan as an essential strategy for improving cooperation, synergies and coherence among member organizations of the Partnership. Members of the Forum are also encouraged to provide voluntary financial contributions to support the activities of the Partnership and its member organizations.

United Nations system

41. Several United Nations bodies, organizations and specialized agencies not participating in the Collaborative Partnership on Forests address issues that are relevant to forests, such as eradication of poverty in its all forms, gender equality and the empowerment of women, labour standards, small and medium-sized enterprises, scientific and technical cooperation, disaster risk reduction, ecotourism and issues related to *the United Nations Declaration on the Rights of Indigenous Peoples*. These bodies, organizations and specialized agencies, within the scope of their mandates, are invited to use the strategic plan as a reference, with a view to building synergies between the global forest goals and targets of the strategic plan and their respective policies and programmes, including their contributions to the achievement of the Sustainable Development Goals.

42. Close cooperation with the secretariats of, and the parties to, the *Rio conventions*[①], and mutually supportive implementation of their forest-related objectives, is important to achieve the global forest goals and targets.

43. The United Nations System Chief Executives Board for Coordination is also invited to promote the use of the strategic plan as a reference for forest-related work within the United Nations system, where appropriate.

---

① *Convention on Biological Diversity, United Nations Convention to Combat Desertification in Those Countries Experiencing Serious Drought and/or Desertification, Particularly in Africa, and United Nations Framework Convention on Climate Change.*

Other intergovernmental partners and stakeholders at the international level

44. In addition to under the multilateral environmental agreements that are represented in the Collaborative Partnership on Forests, forest-related activities are undertaken under a number of other multilateral environmental agreements, such as the *Convention on Wetlands of International Importance especially as Waterfowl Habitat*[①], the *Convention on International Trade in Endangered Species of Wild Fauna and Flora*[②] and the *Convention Concerning the Protection of the World Cultural and Natural Heritage*, and can make important contributions to the global forest goals and targets. The secretariats of and parties to these agreements are invited to seek opportunities to contribute to the implementation of the strategic plan, where appropriate and consistent with their mandates.

Regional and subregional organizations and processes

45. Regional bodies, notably the United Nations regional economic commissions and the FAO regional forestry commissions, and other regional and subregional bodies and processes provide a crucial bridge between international policies and national actions and are important partners in efforts to implement the strategic plan and achieve its global forest goals and targets.

46. The Forum works with regional and subregional bodies and processes to identify ways to contribute to the global forest goals and targets, including by encouraging them to exchange information, enhance cooperation, raise awareness, strengthen stakeholder engagement and build capacity to scale up best practices within and across regions.

47. Regional and subregional bodies and processes, including those within the United Nations system, as well as the criteria and indicator processes, are encouraged to build and strengthen synergies between the strategic plan and their policies and programmes, including in the context of their contributions to the implementation of the Sustainable Development Goals.

48. Members are invited to consider strengthening regional and subregional forest policy development, dialogue and coordination to advance the strategic plan.

Major groups and other stakeholders

49. The effective implementation of sustainable forest management depends on the contributions of all relevant stakeholders, including forest owners, indigenous peoples, local communities, local authorities, the private sector (including small, medium and large forest-based enterprises), non-governmental organizations, women, children, youth, and scientific, academic and philanthropic organizations at all levels.

50. The Forum endeavours to work with major groups and other relevant stakeholders to identify ways to enhance their contributions to the achievement of the global forest goals and targets at all levels and their interactions with the Forum and the Collaborative Partnership on

---

① United Nations, *Treaty Series*, vol. 996, No. 14583.
② Ibid., vol. 993, No. 14537.

Forests, including through networks, advisory groups and other mechanisms, to raise awareness, foster information exchange and dissemination and facilitate coordinated inputs.

51. Major groups and other relevant stakeholders such as private philanthropic organizations, educational and academic entities, volunteer groups and others, are encouraged to autonomously establish and maintain effective coordination mechanisms at all levels for interaction and participation in the Forum and other forest-related United Nations bodies.

B. Means of implementation

52. Building on the *Addis Ababa Action Agenda*, which is an integral part of *the 2030 Agenda for Sustainable Development*, the strategic plan provides a reference for international cooperation and means of implementation, including finance and capacity-building related to forests, supported by effective institutions, sound policies, legal frameworks, good governance and partnerships at all levels. Recognizing that there is no single solution to address all of the needs in terms of financing for actions to achieve the global forest goals and targets, a combination of actions is required at all levels to mobilize resources, by all stakeholders and from all sources, public and private, domestic and international, bilateral and multilateral.

53. The means of implementation targets under Goal 17 and under each Sustainable Development Goal are key to realizing *the 2030 Agenda for Sustainable Development* and are of equal importance with the other Goals and targets. The Agenda, including the Sustainable Development Goals, can be met within the framework of a revitalized Global Partnership for Sustainable Development, supported by the concrete policies and actions as outlined in the *Addis Ababa Action Agenda*. Welcoming the endorsement by the General Assembly of the *Addis Ababa Action Agenda*, which is an integral part of *the 2030 Agenda for Sustainable Development*, it is recognized that full implementation of the Agenda is critical for the realization of the Sustainable Development Goals and targets.

54. Mobilization of and effective use of financial resources, including new and additional resources from all sources and at all levels, is crucial. Public policies and the mobilization and effective use of domestic resources, underscored by the principle of national ownership and leadership, are central to our common pursuit of sustainable development.

55. Private business activity, investment and innovation are major drivers of productivity, inclusive economic growth and job creation, and private international capital flows, particularly foreign direct investment, along with a stable international system, are vital complements to national development efforts.

56. International public finance plays an important role in complementing the efforts of countries to mobilize public resources domestically, especially those with the greatest needs and the least ability to mobilize other resources. An important use of international public finance, including official development assistance, is to catalyse additional resource mobilization from other sources, public and private.

57. Philanthropic organizations and foundations, as well as public-private and multi-stakeholder partnerships, also play important roles in the scaling up of resources for

sustainable forest management and the implementation of the United Nations strategic plan for forests.

58. In advancing the global forest goals and targets, members are encouraged to:

(a) Enhance North-South, South-South and triangular regional and international cooperation on and access to science, technology and innovation and enhance knowledge-sharing on mutually agreed terms, including through improved coordination among existing mechanisms, in particular at the United Nations level, and through a global technology facilitation mechanism;

(b) Promote the development, transfer, dissemination and diffusion of environmentally sound technologies to developing countries on favourable terms, including on concessional and preferential terms, as mutually agreed;

(c) Make full use of the grant and concessional and innovative resources available to them through United Nations system programmes, funds and specialized agencies; national funds and domestic resources; private funding; multilateral, regional and subregional development banks and funding institutions; bilateral development assistance agencies; and funding opportunities provided through foundations and philanthropic organizations.

59. Eligible countries are encouraged to make full use of the international resources available, including through:

(a) The Global Environment Facility (GEF) strategy for sustainable forest management and the GEF focal areas on biodiversity, climate change and land degradation, which serve as funding mechanisms for the *Rio Conventions*;

(b) The GEF strategy and financing for sustainable forest management under the replenishment processes of GEF, including through harnessing synergies across the focal areas of GEF in order to reinforce the importance of sustainable forest management for integrating environmental and development aspirations;

(c) The United Nations Collaborative Programme on Reducing Emissions from Deforestation and Forest Degradation in Developing Countries, activities under the Forest Carbon Partnership Facility and the Forest Investment Programme, and the Green Climate Fund.

60. Members are invited to make full use of the potential of innovative funding mechanisms, including payment for ecosystem services schemes and existing mechanisms under the *United Nations Framework Convention on Climate Change*.

61. Effective attainment of the global forest goals and targets also requires the mobilization of the best available scientific and traditional forest-related knowledge. The scientific community, in cooperation with the Forum and its members, is encouraged to support the implementation of the strategic plan, through scientific inputs presented to the Forum. In doing so, the Forum is invited to build upon existing joint initiatives of the Collaborative Partnership on Forests and further strengthen these initiatives.

Global Forest Financing Facilitation Network

62. The Global Forest Financing Facilitation Network of the United Nations Forum on

Forests, in close cooperation with members of the Collaborative Partnership on Forests, contributes to the scaling up of sustainable forest management by facilitating access by countries to resources to implement the strategic plan and to achieve its global forest goals and targets. To this end, the priorities for the Network are:

(a) To promote and assist members in designing national forest financing strategies to mobilize resources for sustainable forest management, including existing national initiatives, within the framework of national forest programmes or other appropriate national frameworks;

(b) To assist countries in mobilizing, accessing and enhancing the effective use of existing financial resources from all sources for sustainable forest management, taking into account national policies and strategies;

(c) To serve as a clearing house and database on existing, new and emerging financing opportunities and as a tool for sharing lessons learned and best practices from successful projects, building on the Collaborative Partnership on Forests online sourcebook for forest financing;

(d) To contribute to the achievement of the global forest goals and targets, as well as priorities contained in the quadrennial programmes of work.

63. Special consideration should be given to the special needs and circumstances of Africa, the least developed countries, countries with low forest cover, countries with high forest cover, countries with medium forest cover and low deforestation, and small island developing States, as well as countries with economies in transition, in gaining access to funds.

Trust Fund for the United Nations Forum on Forests

64. The Trust Fund for the United Nations Forum on Forests was established in 2001 to finance activities in support of the Forum from voluntary extrabudgetary resources to support its activities. The Trust Fund can be used to support the activities of the Global Forest Financing Facilitation Network. Members of the Forum are encouraged to provide voluntary contributions to the Trust Fund.

65. The Forum is to monitor and assess the work and performance of the Global Forest Financing Facilitation Network on a regular basis, including the availability of Trust Fund resources.

IV. Review Framework

A. Review of the international arrangement on forests

66. In accordance with section XII of Economic and Social Council resolution 2015/33, in 2024 the Forum is to conduct a midterm review of the effectiveness of the international arrangement on forests in achieving its objectives, and a final review in 2030. In the context of the midterm review, the Forum could consider:

(a) A full range of options, including the adoption of a legally binding instrument on all types of forests, the strengthening of the current arrangement and the continuation of the current arrangement;

(b) A full range of financing options, inter alia, the establishment of a voluntary global forest fund in order to mobilize resources from all sources in support of the sustainable management of all types of forests. This can be further considered, if there is a consensus to do so, at a session of the Forum prior to 2024.

B. Progress in implementing the United Nations strategic plan for forests

67. The Forum should assess progress in implementing the United Nations strategic plan for forests in the context of its midterm and final reviews of the effectiveness of the international arrangement on forests, in 2024 and 2030. The assessment should be based on internationally agreed indicators, including relevant Sustainable Development Goal indicators, that are relevant to the global forest goals and targets.

68. The assessment should take into account voluntary national reporting on the implementation of the United Nations strategic plan for forests, the United Nations forest instrument, voluntary national contributions and the results of the most recent Global Forest Resources Assessment of FAO, as well as inputs from the Collaborative Partnership on Forests and its member organizations and other partners within and outside of the United Nations system, including regional and subregional organizations and relevant stakeholders.

69. To reduce the reporting burden, the Forum is to establish a cycle and format for voluntary national reporting by its members, taking into account the cycle of the Global Forest Resources Assessments and the Sustainable Development Goal review cycle at the global level.

C. Contributing to the follow-up to and review of *the 2030 Agenda for Sustainable Development*

70. The United Nations Forum on Forests, as a functional commission of the Economic and Social Council, should contribute to the follow-up to and review of *the 2030 Agenda* and its Sustainable Development Goals, including through the work of the Collaborative Partnership on Forests on global forest indicators, as well as highlight the contribution of forests to the Sustainable Development Goals, to be reviewed in depth at the annual sessions of the high-level political forum on sustainable development.

V. Communication and Outreach Strategy

71. Communication and outreach are essential components of the United Nations strategic plan for forests, which is itself an important communication tool. A communication and outreach strategy should be developed to raise awareness, within and outside of the forest sector, of the vital contribution of all types of forests and trees to life on earth and human well-being. The communications and outreach strategy should draw on the strategic plan, synchronize with the quadrennial programmes of work and consider relevant themes, including those which are relevant to the high-level political forum on sustainable development. Actors at all levels are encouraged to contribute to these efforts.

72. The communication strategy should raise the awareness of the United Nations strategic plan for forests, including through its graphic visualization.

73. The Forum should develop the communication and outreach strategy for the strategic plan. This strategy should identify, inter alia, target audiences, key messages, methods,

activities and success criteria.

74. The United Nations system, the Collaborative Partnership on Forests and its member organizations and other partners are encouraged to enhance cooperation and synergies on forest-related communications and outreach to increase the impact of their messaging, and to consider joint events and products with national, regional, subregional and non-governmental organizations and processes.

75. The International Day of Forests on 21 March provides a powerful annual event to promote implementation of the strategic plan, and is a platform for individual and collective public outreach activities by all actors on forests at all levels. Members are encouraged to observe this day in collaboration with other relevant stakeholders and to inform the Forum about their activities.

Notes

1. *Convention on International Trade in Endangered Species of Wild Fauna and Flora*

《濒危野生动植物种国际贸易公约》于 1973 年在美国华盛顿签署，所以也称《华盛顿公约》。其宗旨是通过物种分类和许可证制度控制野生动植物种及其产品的国际贸易，保护濒危野生动植物种不会因为国际贸易而遭到过度开发利用。

2. the *Aichi Biodiversity Targets*

2010 年，《生物多样性公约》缔约国大会第十次会议在日本爱知县举办，会上通过了 2011—2020 年《生物多样性战略计划》（简称《战略》），《战略》中的 5 个战略目标及相关的 20 个纲要目标统称为"爱知生物多样性目标"（简称"爱知目标"）。爱知目标旨在激励所有国家和利益相关方采取措施，推动实现物种多样性、基因多样性（或称遗传多样性）和生态多样性三大生物多样性目标。

3. the *Addis Ababa Action Agenda*

2015 年 7 月，联合国第三次发展筹资问题国际会议上达成了《亚的斯亚贝巴行动议程》。该议程包含 100 多个具体措施，用于支持可持续发展目标的落实，并且在科技、基础设施、社会保障、卫生、中小微企业、外国援助、税收、气候变化以及针对最贫困国家的一揽子援助措施方面提出了新的举措。

4. The Global Environment Facility

全球环境基金（GEF）是世界银行 1990 年创建的实验项目，其宗旨是与国际机构、社会团体及私营部门合作，支持环境友好工程，协力解决环境问题。

5. *United Nations Declaration on the Rights of Indigenous Peoples*

2007 年 9 月，第 61 届联合国大会通过了《土著人民权利宣言》。该宣言保障土著人不受歧视，并强调保障土著人民保持和加强自身制度、文化和传统的权利，以及拥有按照自身需要和愿望选择发展道路的权利。

Key Words and Phrases

1. ambitious	/æmˈbɪʃəs/	adj.	having a desire to achieve a particular goal 有志向的；有抱负的
2. amend	/əˈmend/	v.	to change the law or text in order to it improve or make it more accurate 修正
3. appendix	/əˈpendɪks/	n.	extra information that is placed after the end of the main text 附录
4. diffusion	/dɪˈfjuːʒn/	n.	the spread of cultural elements from one area or group of people to others by contact 传播；扩散
5. encompass	/ɪnˈkʌmpəs/	v.	to include a large number of things 包含
6. fodder	/ˈfɒdə(r)/	n.	food that is given to cows, horses and other animals 饲料
7. holistic	/həˈlɪstɪk/	adj.	relating to or concerned with wholes or with complete systems rather than with the analysis of parts 整体的
8. incentive	/ɪnˈsentɪv/	n.	something that incites or has a tendency to incite to action 动机；刺激
9. indicative	/ɪnˈdɪkətɪv/	adj.	showing or suggesting sth. 象征的
10. integral	/ˈɪntɪɡrəl/	adj.	essential to completeness 必需的；整体的
11. interplay	/ˈɪntəpleɪ/	n.	the way in which two or more things or people affect each other 相互影响
12. notably	/ˈnəʊtəbli/	adv.	especially, in particular 尤其；特别
13. proclaim	/prəˈkleɪm/	v.	to formally make something known to the public 宣告，声明
14. underscore	/ˌʌndəˈskɔː(r)/	v.	to emphasize or show that something is important or true 强调
15. visualization	/ˌvɪʒuəlaɪˈzeɪʃn/	n.	the act or process of forming a mental image 想象

16. alien species　　外来物种
17. forest landscape rehabilitation　　森林景观恢复
18. eradication of poverty　　消除贫困
19. non-exhaustive list　　非全面的清单
20. on concessional and preferential terms　　以优惠的条件
21. per capita　　人均
22. scale up　　增大
23. thematic area　　专题领域

Exercises

Exercise 1 Reading Comprehension

Directions: Read Ⅲ. *Implementation Framework B. Means of implementation (52-61)* of **United Nations Strategic Plan for Forests 2017—2030** *and decide whether the following statements are true or false. Write T for true or F for false in the brackets in front of each statement.*

1. (　　) Stakeholders and all sources including public and private, domestic and international should work together to mobilize resources to achieve the global forest goals and targets.

2. (　　) Public policies and mobilization and effective use of resources from all sources especially international resources are central to the common pursuit of sustainable development.

3. (　　) Philanthropic organizations are the only driving forces to scale up resources for sustainable forest management and the implementation of the strategic plan for forests.

4. (　　) Members are encouraged to enhance North-South, South-South cooperation on and access to science, technology and innovation and enhance knowledge-sharing through a global technology facilitation mechanism.

5. (　　) Members are encouraged to promote the transfer and diffusion of environmentally sound technologies to stakeholders on favorable terms.

6. (　　) The scientific community should work together with the Forum and make full use of scientific inputs to support the implementation of the strategic plan.

Exercise 2 Skimming and Scanning

Directions: *Read the following passage excerpted from* **United Nations Strategic Plan for Forests 2017—2030**. *At the end of the passage, there are six statements. Each statement contains information given in one of the paragraphs of the passage. Identify the paragraph from which the information is derived. Each paragraph is marked with a letter. You may choose a paragraph more than once. Answer the questions by writing the corresponding letter in the brackets in front of each statement.*

A) At the heart of the strategic plan are six global forest goals and 26 associated targets to be achieved by 2030. The global forest goals and targets are voluntary and universal. They support the objectives of the international arrangement on forests and are aimed at contributing to progress on the Sustainable Development Goals and other international forest-related instruments, processes, commitments and goals.

B) The vision, principles and commitments set out in *the 2030 Agenda for Sustainable Development* provide the context for the global forest goals and targets, which are interconnected and integrate the economic, social and environmental dimensions of sustainable forest management and sustainable development.

C) The global forest goals and targets are intended to stimulate and provide a framework for voluntary actions, contributions and enhanced cooperation by countries and international, regional, subregional and non-governmental partners and stakeholders. They also provide a reference for enhanced coherence and collaboration on forests within the United Nations system and among member organizations of the Collaborative Partnership on Forests, as well as among other forest-related organizations and processes.

D) The global goals and targets encompass a wide variety of thematic areas in regard to which voluntary actions, contributions and cooperation are needed to advance their achievement. These thematic areas reflect and encompass the 44 policies, measures and actions set out in *the United Nations Forest Instrument*. A non-exhaustive list of indicative thematic areas for action is contained in the appendix to the present document. Indicative thematic areas may correspond to more than one goal.

Global forest goal 1

E) Reverse the loss of forest cover worldwide through sustainable forest management, including protection, restoration, afforestation and reforestation, and increase efforts to prevent forest degradation and contribute to the global effort of addressing climate change. By 2020, promote the implementation of sustainable management of all types of forests, halt deforestation, restore degraded forests and substantially increase afforestation and reforestation globally.

Global forest goal 2

F) Enhance forest-based economic, social and environmental benefits, including by improving the livelihoods of forest-dependent people. Increase the access of small-scale forest enterprises, in particular in developing countries, to financial services, including affordable credit, and their integration into value chains and markets.

Global forest goal 3

G) Increase significantly the area of protected forests worldwide and other areas of sustainably managed forests, as well as the proportion of forest products from sustainably managed forests. The area of forests worldwide designated as protected areas or conserved through other effective area-based conservation measures is significantly increased.

Global forest goal 4

H) Mobilize significantly increased, new and additional financial resources from all sources for the implementation of sustainable forest management and strengthen scientific and technical cooperation and partnerships.

Global forest goal 5

I) Promote governance frameworks to implement sustainable forest management, including through the United Nations forest instrument, and enhance the contribution of forests to *the 2030 Agenda for Sustainable Development*.

Global forest goal 6

J) Enhance cooperation, coordination, coherence and synergies on forest-related issues at all levels, including within the United Nations system and across member organizations of the Collaborative Partnership on Forests, as well as across sectors and relevant stakeholders.

1. (　　) Sustainable forest management should be done to reverse the loss of forest cover and contribute to the global effort of addressing climate change.

2. (　　) United Nations system and other member organizations should cooperate and coordinate on forest-related issues at all levels.

3. (　　) The global forest goals and targets are aimed at contributing to progress on the Sustainable Development Goals and other international forest-related instruments, commitments and goals.

4. (　　) The area of forests worldwide designated as protected areas or conserved is increased through effective area-based conservation measures.

5. (　　) The global goals and targets cover various thematic areas which need voluntary actions, contributions and cooperation to advance their achievement.

6. (　　) Small-scale forest enterprises in developing countries should be offered more access to financial services, including affordable credit and their integration into value chains and markets.

Exercise 3　Word Formation

Directions: *In this section, there are ten sentences from **United Nations Strategic Plan for Forests 2017—2030**. You are required to complete these sentences with the proper form of the words given in blanks.*

1. United Nations Forum on Forests should develop a concise plan to serve as a _____ framework to enhance the coherence of and guide and focus the work of the international arrangement on forests and its components. (strategy)

2. Continued rapid population growth, as well as rising per capita income, is accelerating the global demand for and _____ of forest products and services and putting pressure on forests. (consume)

3. The effective implementation of sustainable forest management is _____ dependent upon adequate resources, including financing, capacity development and the transfer of environmentally sound technologies. (critical)

4. One of the targets is to increase the access of small-scale forest enterprises, in particular in developing countries, to financial services, including _____ credit, and their integration into value chains and markets. (afford)

5. The contribution of all types of forests to biodiversity conservation and climate change _____ and adaptation is enhanced, taking into account the mandates and ongoing work of relevant conventions and instruments. (mitigate)

6. Forest-related programmes within the United Nations system are coherent and _____ and integrate the global forest goals and targets, where appropriate. (complement)

7. By 2020, efforts are made to promote the implementation of sustainable management of all types of forests, halt deforestation, restore degraded forests and _____ increase afforestation and reforestation globally. (substantial)

8. The individual and collective actions and commitments of members of the Forum are

_____ for the successful implementation of the strategic plan and achievement of its global forest goals and targets. (decision)

9. Several United Nations bodies address issues that are relevant to forests, such as eradication of poverty in its all forms, gender equality and the _____ of women and scientific and technical cooperation. (power)

10. The secretariats of and parties to these agreements are invited to seek opportunities to contribute to the _____ of the strategic plan, where appropriate and consistent with their mandates. (implement)

Exercise 4 Translation
Section A
Directions: Read *United Nations Strategic Plan for Forests 2017—2030*, and complete the sentences by translating into English the Chinese given in blanks.

1. The strategic plan _____ (为开展涉林工作提供参考构架) of the United Nations system and for the fostering of enhanced coherence, collaboration and synergies among United Nations bodies and partners towards the vision and mission set out below.

2. Forests are home to an estimated 80% of all terrestrial species. Forests contribute substantially to _____ (减缓与适应气候变化和保护生物多样性).

3. As set out in the United Nations forest instrument, "Sustainable forest management, as _____ (一个动态和不断发展的概念), is intended to maintain and enhance the economic, social and environmental values of all types of forests, for the benefit of present and future generations".

4. The number of countries that have _____ (制定和实施林业投融资战略) and have access to financing from all sources is significantly increased.

5. Members may include in their voluntary national contributions, as appropriate, the forest-related contributions they intend to make with regard to _____ (其他国际涉林承诺和目标), such as the implementation of *the 2030 Agenda* and its Sustainable Development Goals, the Aichi Biodiversity Targets and actions to address climate change under *the Paris Agreement*.

6. The Forum endeavours to work with major groups and other relevant stakeholders to identify ways to enhance their contributions to _____ (实现全球森林目标) at all levels.

7. _____ (资金的筹集和有效利用) including new and additional resources from all sources and at all levels, is crucial. Public policies underscored by the principle of national ownership and leadership are central to our common pursuit of sustainable development.

8. In advancing the global forest goals and targets, members are encouraged to promote the development, transfer, dissemination and diffusion of environmentally sound technologies to developing countries on favorable terms, including _____ (以共同商定

的优惠条件).

9. Members are invited to make full use of the potential of innovative funding mechanisms, including _____ (生态系统服务价值补偿机制) and existing mechanisms under the United Nations Framework Convention on Climate Change.

10. The Forum is to _____ (监测和评估工作和运行情况) of the Global Forest Financing Facilitation Network on a regular basis, including the availability of Trust Fund resources.

Section B

Directions: *Translate the following sentences from English into Chinese.*

1. A communication and outreach strategy should be developed to raise awareness, within and outside of the forest sector, of the vital contribution of all types of forests and trees to life on earth and human well-being. The communications and outreach strategy should draw on the strategic plan, synchronize with the quadrennial programmes of work and consider relevant themes, including those which are relevant to the high-level political forum on sustainable development. (*Communication and Outreach Strategy*)

2. The Forum should assess progress in implementing the United Nations strategic plan for forests in the context of its midterm and final reviews of the effectiveness of the international arrangement on forests, in 2024 and 2030. The assessment should be based on internationally agreed indicators, including relevant Sustainable Development Goal indicators, that are relevant to the global forest goals and targets. (*Progress in Implementing the United Nations Strategic Plan for Forests*)

3. International public finance plays an important role in complementing the efforts of countries to mobilize public resources domestically, especially those with the greatest needs and the least ability to mobilize other resources. An important use of international public finance, including official development assistance, is to catalyse additional resource mobilization from other sources, public and private. (*Implementation Framwork*)

4. The global goals and targets encompass a wide variety of thematic areas in regard to which voluntary actions, contributions and cooperation are needed to advance their achievement. These thematic areas reflect and encompass the 44 policies, measures and actions set out in *the United Nations Forest Instrument*. A non-exhaustive list of indicative thematic areas for action is contained in the appendix to the present document. Indicative thematic areas may correspond to more than one goal. (*Global Forests Goals and Targets*)

5. Forests are among the world's most productive land-based ecosystems and are essential to life on earth. The United Nations strategic plan for forests 2017—2030 provides a global framework for action at all levels to sustainably manage all types of forests and trees outside forests, and to halt deforestation and forest degradation. The strategic plan also provides a framework for forest-related contributions to the implementation of *the 2030 Agenda for Sustainable Development*, *the United Nations Forest Instrument* and other international forest-related instruments, processes, commitments and goals. (*Introduction*)

6. The effective implementation of sustainable forest management is critically dependent upon adequate resources, including financing, capacity development and the transfer of environmentally sound technologies and, in particular, the need to mobilize increased financial resources, including from innovative sources, for developing countries, including least developed countries, landlocked developing countries and small island developing States, as well as countries with economies in transition. (*Shared llnited Natiogs Vision*)

Extensive Readings

Passage 1

Directions: *Read the following passage and choose the best answer for each of the following questions according to the information given in the passage.*

"Drastically reducing deforestation and systemically restoring forests and other ecosystems is the single largest nature-based opportunity for climate mitigation."—UN Secretary General António Guterres speaking on the State of the Planet Global Forest Goal 1 (GFG1) calls for reversing the loss of forest cover worldwide through sustainable forest management. Forests currently cover 31% of the global land area. Between 2010 and 2020, global forest area fell by 1.2%, with declines concentrated in Africa and South America. However, within this global trend, and since 1990, Asia, Europe, and Oceania saw net increases in forest area: the forest area of this group of regions increased by 1.1% between 2010 and 2020. Further, according to the *FRA 2020*, "The rate of net forest loss decreased substantially over the period 1990—2020 due to a reduction in deforestation in some countries, plus increases in forest area in others through afforestation and the natural expansion of forests." Between 2015 and 2020, deforestation, which measures the conversion of forest to other land use, stood at 10.2 million hectares (ha) per year. This was rather less than in earlier periods. Within this same five-year window, total forest expansion by afforestation or natural expansion was 4.7 million ha per year, with Asia registering the largest expansion. Forest ecosystems are the largest terrestrial carbon sink, absorbing roughly 2 billion tonnes of CO_2 each year. Between 1990 and 2010, the total global forest carbon stock fell from 668 gigatonnes (Gt) in 1990 to 662 Gt in 2010, mainly due to a loss of forest area. In 2020, it stayed at 662 Gt, with Europe, North and Central America, and South America housing two thirds of this total. The global carbon stock comprised approximately 300 Gt of soil organic matter, 275 Gt of living biomass, and 88 Gt of dead wood and litter.

Forest area is increased by 3% worldwide. Countries developed strategies and plans to maintain or increase their forest area, often with quantified targets, detailed objectives, methods, and resources. Sometimes these strategies covered only the forest sector, such as national forest programmes. However, in other instances, plans were part of broader national strategies, such as for mitigating climate change or addressing national development and poverty reduction goals. Some countries increased the availability of resources to expand forest area or reduce deforestation. China and Liberia, for instance, drafted clear guidelines for silviculture and afforestation. These guidelines were met with the provision of training for all relevant activities and support for research and technical assistance, for example on tree breeding and seedling production. Some countries developed partnerships with industry, including in sectors beyond forestry, to prevent the loss of forest area. Other countries organized tree planting programmes and events, often with the participation of civil society and the general public. On National Tree Day in the Central African Republic, the government invited each citizen to plant at least one seedling provided by the Forest Service.

The world's forest carbon stocks are maintained or enhanced. The inclusion of forest-related actions in national climate strategies and programmes impacted favorably on the achievement of GFG by raising public visibility of and political will for maintaining and enhancing carbon stocks. Incorporating forest and climate action also opened access to resources and promoted intersectoral approaches to achieving forest related goals and objectives. International programmes, notably REDD+, as well as core work under the UNFCCC played a major role in this regard. International funding sources for climate change mitigation and adaptation included the Green Carbon Fund, the Forest Carbon Partnership Facility (FCPF), and the Climate Investment Fund. Monitoring and reporting of carbon stocks improved. These often progressed according to standard international methods, notably greenhouse gas (GHG) inventories and national reference levels, as was the case in Madagascar, the Philippines, the Republic of Korea, and Turkey. In some countries, improved national forest inventories accompanied strengthened monitoring and reporting. Countries also innovated. The Republic of Korea, for instance, implemented a forest carbon offset scheme, while New Zealand introduced changes to strengthen its Emission Trading Scheme by increasing economic incentives for afforestation, as administered by Forestry New Zealand. United States considered a carbon tax to curb the exploitation of wood that led to deforestation. Germany developed its REDD+ Action Plan "Forests for Good Living", through which the country sought to reduce gross CO_2 emissions by at least 20% by 2025 through policies and measure focused on reducing deforestation. Other countries took measures to ensure that harvests remained at sustainable levels.

The year 2020 saw record fire seasons in Australia and the United States of America, among other countries, and these events contributed to global carbon emissions. Algeria, Bulgaria, Guinea, the Slovak Republic, and the United States of America each reported taking specific measures, such as procuring equipment, carrying out trainings, offering funding, and implementing systems to prevent and control fires and their consequent impacts on GHG emissions. Finally, research helped deepen forest-climate links, most notably in the areas of assessing carbon flows in forest ecosystems and in harvested wood products; improved monitoring methods; and in the construction of scenarios to inform policymaking. Such research set the stage for the continued and additional integration of forest and climate objectives, and the maintenance, if not enhancement, of global forest carbon stocks.

"By 2020, promote the implementation of sustainable management of all types of forests, halt deforestation, restore degraded forests and substantially increase afforestation and reforestation globally." Related to this target, in many cases, countries revised and modified forest laws, codes, and institutions to enhance progress towards sustainable forest management. Policy instruments often explicitly stated the principles underlying sustainable forest management. Overall, a number of countries put in place measures to protect forests, halt deforestation, and expand the forest area legally protected with a view to preserving biodiversity, native forests, and forest ecosystem functions.

1. Which statement is the direct evidence to show the decrease of net forest loss? _____
   A. Deforestation from 2015 to 2020 covered at 10.2 million hectares per year.
   B. Asia registered the largest expansion of natural plants in the world.
   C. The global forest carbon stock fell from 668 Gt in 1990 to 662 Gt in 2010.
   D. Global Forest area declined mainly in Africa and South America.

2. The largest part of global carbon stock is from _____.
   A. living biomass   B. dead wood and litter
   C. soil organic matter   D. trees and plants

3. All the following strategies are implemented to increase forest area EXCEPT _____.
   A. developing national forest programs
   B. developing partnership with industry
   C. incorporating forest plans into poverty reduction goals
   D. improving the public awareness of environmental protection

4. Which of the following is NOT mentioned as an innovative measure taken by countries to enhance forest carbon stocks? _____
   A. Increasing economic incentives for afforestation.
   B. Monitoring and reporting carbon stocks.
   C. Imposing a carbon tax on wood cutting.
   D. Implementing a forest carbon offset scheme.

5. In 2020, Australia and the United States contributed to global carbon emissions due to _____.
   A. deforestation   B. urbanization
   C. severe wild fires   D. extreme climatic events

Passage 2

Directions: *In this section, there is a passage with twelve blanks. You are required to select one word for each blank from a list of choices given in a word bank following the passage. Read the passage through carefully before making your choices. Each choice in the bank is identified by a letter. You may not use any of the words in the bank more than once.*

A. ecosystems	B. attributed	C. concentration	D. competing
E. genetic	F. enhance	G. maximum	H. impacts
I. collectively	J. yield	K. emitting	L. associated

Forests are resilient ecosystems, there are limits to their ability to withstand environmental change, and they degrade when these limits are exceeded. Understanding these limits also allows us to ____1____ various forest outputs through silviculture. There are many examples of thinning, selective harvesting and manipulation of watershed forests to increase the ____2____ of wood, water and wildlife without any apparent negative ecological impact. On the other hand, some industrial forestry activities have been ____3____ with a number of environmental stresses on forest ecosystems. These activities include harvesting; road construction; manipulation of cover types and species through silviculture and deforestation and through the use of mechanical, biological and chemical technologies for protection against fire, insects, diseases and ____4____ vegetation.

Forests are also exposed to environmental stresses associated with other human activities such as industrial manufacturing and the use of fossil fuels. The ____5____ of some of these stresses are restricted and local, others are global. For example, while forest decline in certain parts of Europe is ____6____ to airborne pollutants, all types of the world's forests would be exposed to the anticipated global warming associated with an increase in the ____7____ of greenhouse gases in the atmosphere. A managed natural forest, a forest plantation or an ecological reserve to study natural processes would be of limited value, if any, in the vicinity of a manufacturing facility ____8____ stressful pollutants.

Sustainable forest development, therefore, means recognizing the limits of forests to withstand environmental change, individually and ____9____, and in managing human activities to produce the ____10____ level of benefits obtainable within these limits. Three critical parameters: productive capacity, renewal capacity and spsecies and ecological diversity may be used to assess the status of forests with regard to individual species and ____11____. Productive capacity is a function of the number of species and individual trees growing on a location. Renewal of a forest ecosystem, following harvesting or other forms of disturbances, is dependent on the nature and intensity of disturbance and the mode of reproduction of species located on the site. Species and ecological diversity mean that forests are a rich repository of planet Earth's ____12____ heritage.

Further Studies and Post-Reading Discussion

Task 1
Directions: *Surf the Internet and find more information about* **United Nations Strategic Plan for Forests 2017—2030**. *Work in groups and work out a report on one of the following topics.*
1. Thematic areas for action associated with the global forest goal.
2. The policies and measures taken by China in protecting forests.

Task 2

Directions: *Read the following sentences on Eco-Civilization and make a speech on your understanding of the eco-environmental conservation.*

Building a Beautiful China

走向生态文明新时代，建设美丽中国，是实现中华民族伟大复兴的中国梦的重要内容。中国将按照尊重自然、顺应自然、保护自然的理念，贯彻节约资源和保护环境的基本国策，更加自觉地推动绿色发展、循环发展、低碳发展，把生态文明建设融入经济建设、政治建设、文化建设、社会建设各方面和全过程，形成节约资源、保护环境的空间格局、产业结构、生产方式、生活方式，为子孙后代留下天蓝、地绿、水清的生产生活环境。 保护生态环境，应对气候变化，维护能源资源安全，是全球面临的共同挑战。中国将继续承担应尽的国际义务，同世界各国深入开展生态文明领域的交流合作，推动成果分享，携手共建生态良好的地球美好家园。（摘自《习近平谈治国理政（第二卷）》）	Ushering in a new era of ecological progress and building a beautiful China is an important element of the Chinese Dream. China will respect and protect nature, and accommodate itself to nature's needs. It will remain committed to the basic state policy of conserving resources and protecting the environment. It will promote green, circular and low-carbon development, and promote ecological progress in every aspect of its effort to achieve economic, political, cultural and social progress. China will also develop a resource-efficient and environmentally friendly geographical layout, industrial structure, mode of production and way of life, and leave tour future generations a working and living environment of blue skies, green fields and clean water. Protecting the environment, addressing climate change and securing energy and resources is a common challenge for the whole world. China will continue to assume its due international obligations, carry out in-depth exchanges and cooperation with all other countries in promoting ecological progress, and work with them to promote the sharing of best practices, and make the earth an environmentally sound homeland. (Excerpt from *Xi Jinping: The Governance of China* II)